TAKE NOTE!

TO ACCOMPANY

CHEMISTRY
Matter and Its Changes

Third Edition

James E. Brady
St. John's University, New York

Joel W. Russell
Oakland University, Michigan

John R. Holum
Augsburg College (Emeritus), Minnesota

JOHN WILEY & SONS, INC.
New York · Chichester · Weinheim
Brisbane · Singapore · Toronto

To order books or for customer service call 1-800-CALL-WILEY (225-5945).

Copyright © 2000 John Wiley & Sons, Inc. All rights reserved. No part of this publication may be reproduced, stored in a retrieval system or transmitted in any form or by any means, electronic, mechanical, photocopying, recording, scanning or otherwise, except as permitted under Sections 107 or 108 of the 1976 United States Copyright Act, without either the prior written permission of the Publisher, or authorization through payment of the appropriate per-copy fee to the Copyright Clearance Center, 222 Rosewood Drive, Danvers, MA 01923, (978) 750-8400, fax (978) 750-4470. Requests to the Publisher for permission should be addressed to the Permissions Department, John Wiley & Sons, Inc., 605 Third Avenue, New York, NY 10158-0012, (212) 850-6011, fax (212) 850-6008, E-Mail: PERMREQ@WILEY.COM.

ISBN 0-471-38424-0

Printed in the United States of America

10 9 8 7 6 5 4 3 2 1

Printed and bound by Courier Westford, Inc.

CONTENTS

1	Figure 1.2	33	From Pages 368,	64	Figure 11.17	95	Figure 18.3
2	Figure 1.11		369 & 370	65	Figure 11.19	96	Figure 18.4
3	Figure 2.5	34	Figure 9.2	66	Figure 11.21	97	Figure 18.5
4	Figure 2.10a	35	Figure 9.3	67	Figure 11.24	98	Figure 18.6
5	Figure 2.14	36	Figure 9.4	68	Figure 11.25	99	Figure 18.10
6	Figure 2.15	37	Figure 9.5	69	Figure 11.29	100	Figure 19.2
7	Figure 2.16	38	Figure 9.7	70	Figure 11.30	101	Figure 19.7
8	Figure 2.22	39	Figure 9.8		Figure 11.32	102	Figure 19.8
9	Figure 2.25	40	Figure 9.11	71	Figure 11.33	103	Figure 19.11
10	Figure 4.2	41	Figure 9.12	72	Figure 12.4	104	Figure 19.15
11	Figure 4.3	42	Figure 9.13	73	Figure 12.5	105	Figure 19.16
12	From Page 163	43	Figure 9.16	74	Figure 12.6	106	Figure 20.10
13	From Page 166	44	Figure 9.17	75	Figure 12.9	107	Figure 20.11
14	Figure 6.1	45	Figure 9.20	76	Figure 12.10	108	Figure 20.18
15	Figure 6.3	46	Figure 9.22	77	Figure 12.11	109	Figure 20.21
16	Figure 6.4	47	Figure 9.25	78	Figure 12.13	110	Figure 21.6
17	Figure 6.5	48	Figure 9.28	79	Figure 12.15	111	Figure 21.7
18	Figure 7.3a	49	Figure 9.33	80	Figure 12.17	112	Figure 21.9
19	Figure 7.6	50	Figure 9.34	81	Figure 12.20	113	Figure 21.10
20	Figure 7.15	51	Figure 9.38	82	Figure 12.21	114	Figure 21.16
21	Figure 7.17	52	Figure 10.2	83	Figure 13.4	115	Figure 21.22
22	Figure 7.19	53	Figure 10.10	84	Figure 13.5	116	Figure 22.1
23	Figure 7.20	54	Figure 10.12	85	Figure 13.7	117	Figure 22.8
24	Figure 7.22	55	Figure 10.13	86	Figure 13.8	118	Figure 22.9
25	Figure 7.23	56	Figure 10.14	87	Figure 13.14	119	Figure 22.16
26	Figure 7.25	57	Figure 11.3	88	Figure 13.15	120	Figure 22.17
27	Figure 7.27	58	Figure 11.4	89	Figure 13.16	121	Figure 23.7
28	Figure 7.28	59	Figure 11.6	90	Figure 14.2	122	Figure 23.11
29	Figure 8.1	60	Figure 11.7	91	Figure 15.2	123	Figure 23.12a
30	Figure 8.4	61	Figure 11.8	92	Figure 15.3	124	Figure 23.14
31	Figure 8.6	62	Figure 11.15	93	Figure 16.5	125	Figure 23.15
32	Figure 8.7	63	Figure 11.16	94	Figure 16.6		

FIGURE 1.2

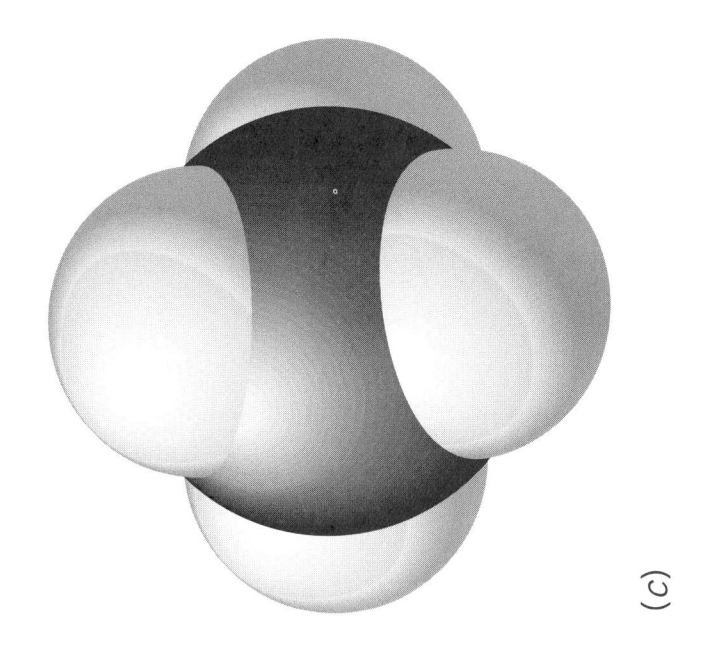

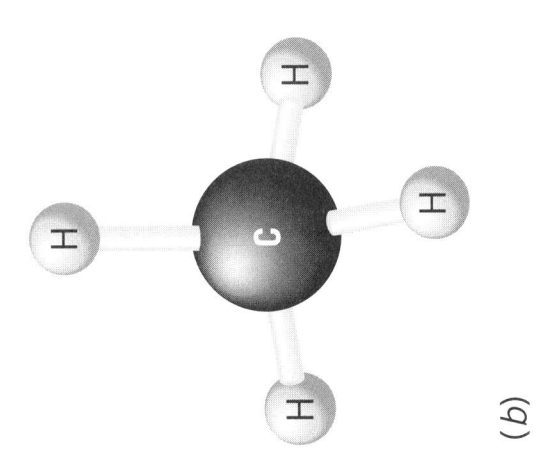

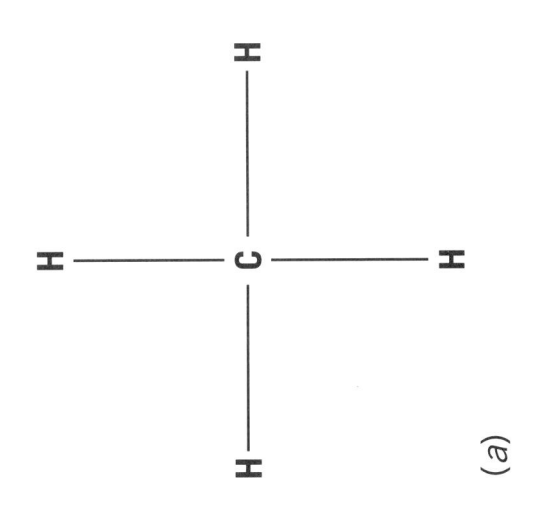

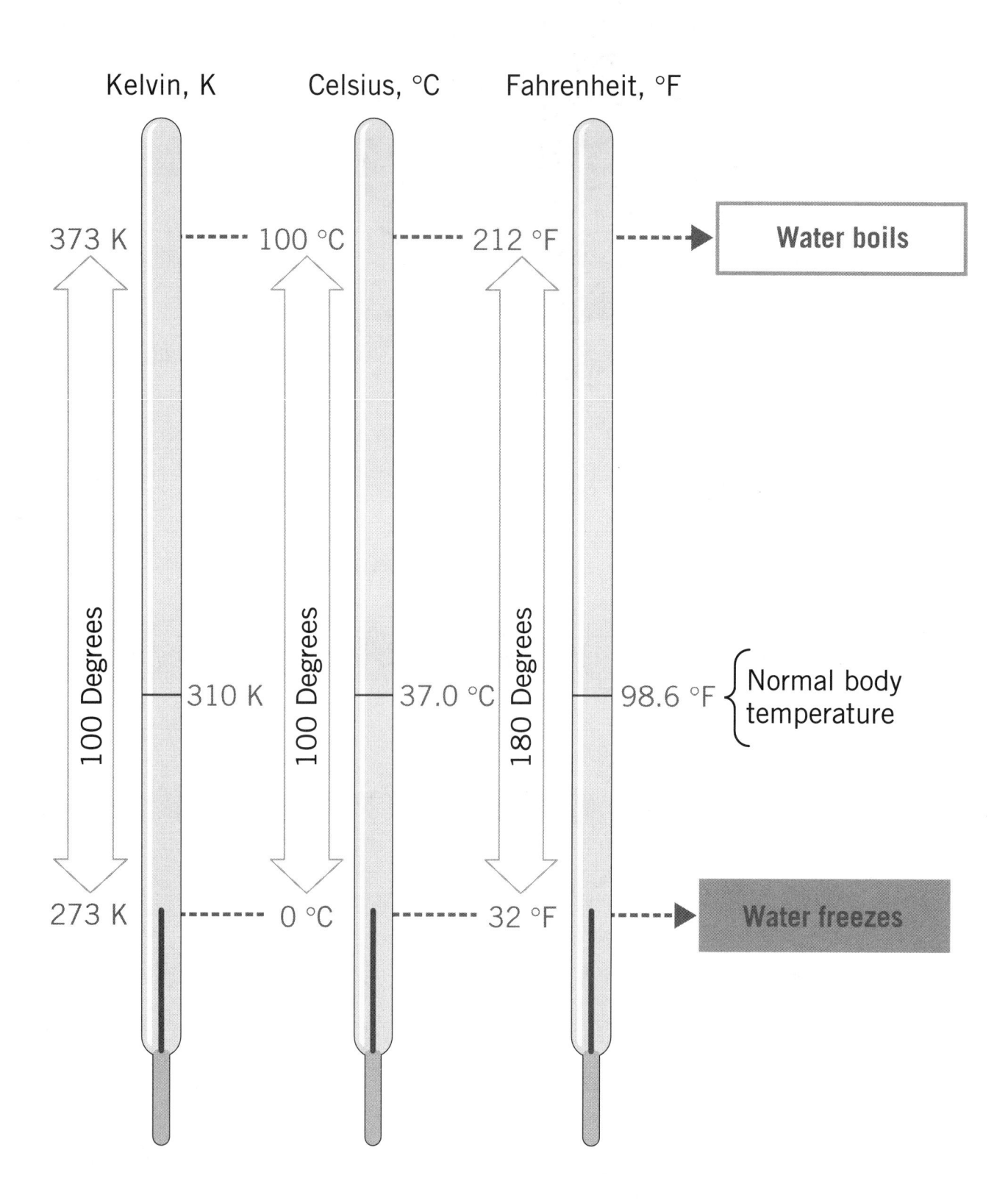

FIGURE 1.11

FIGURE 2.5

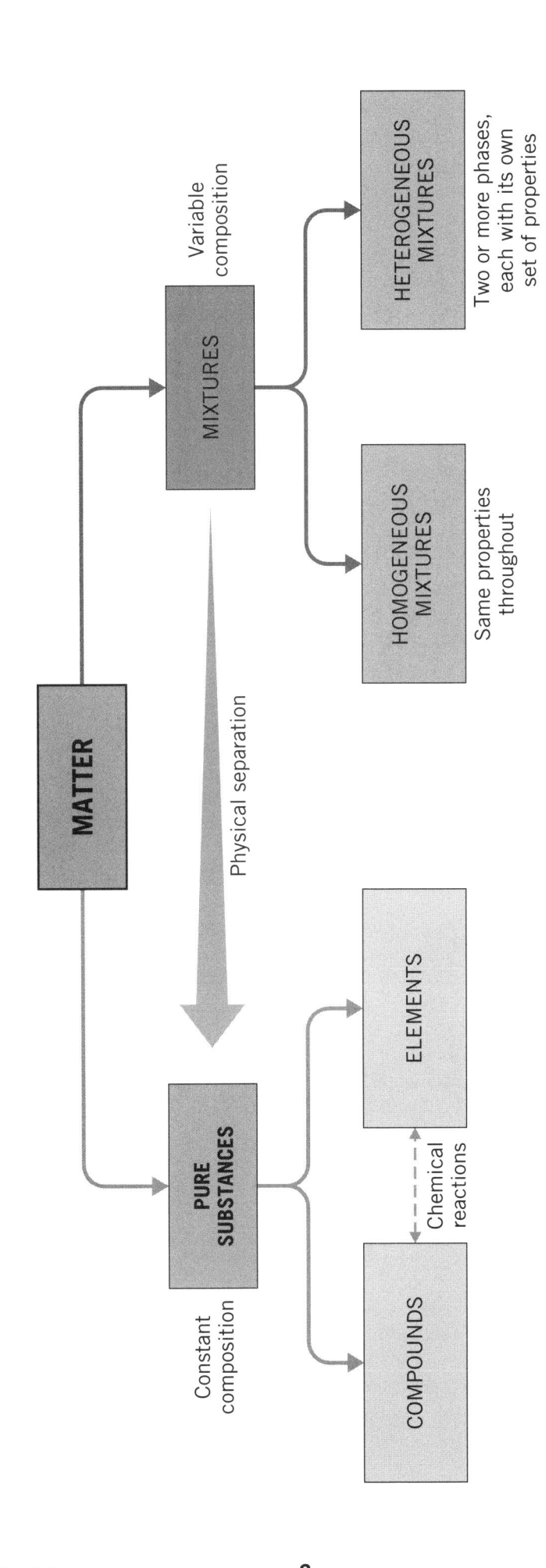

$2H_2 + O_2 \longrightarrow 2H_2O$

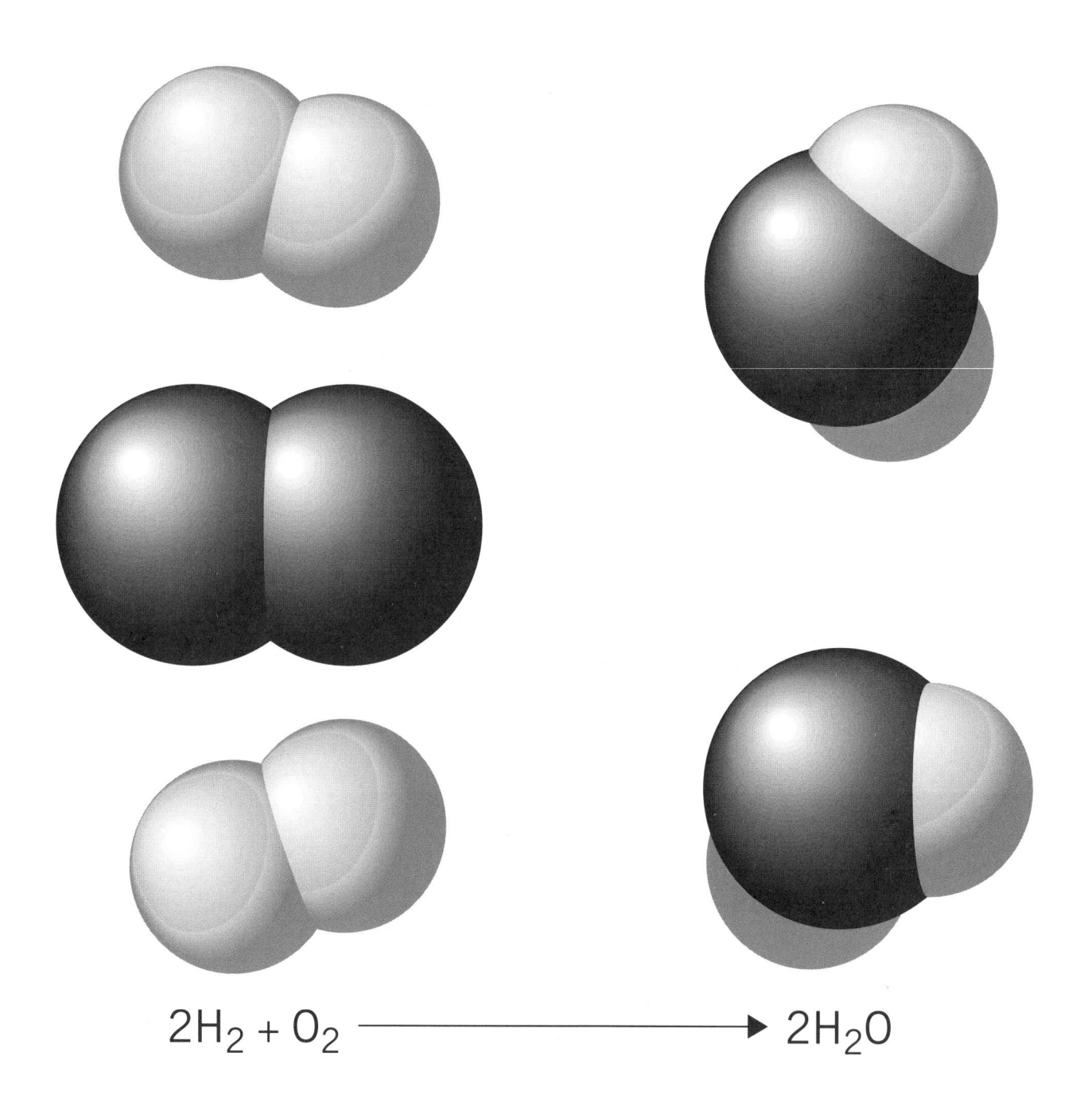

(a)

FIGURE 2.10a

FIGURE 2.14

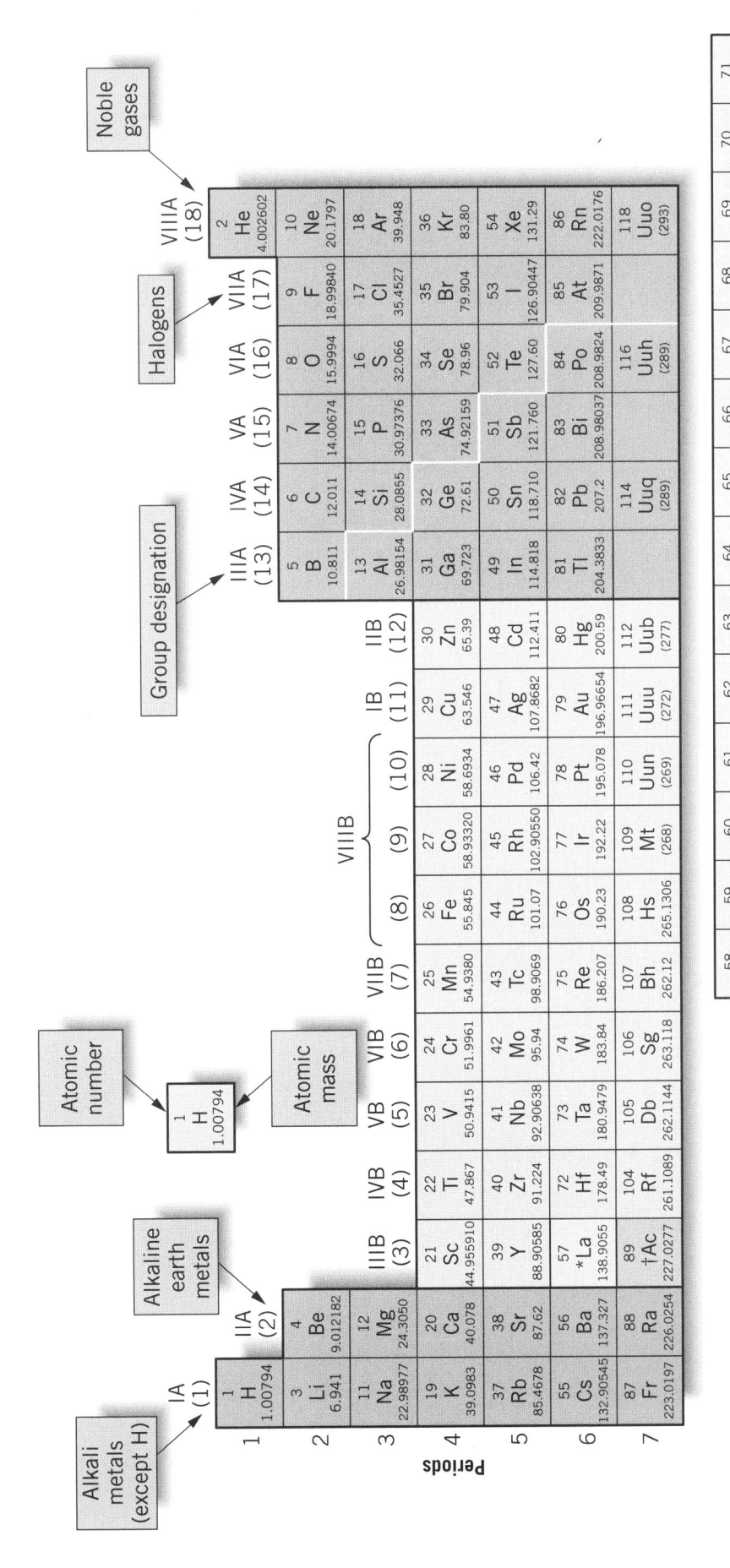

FIGURE 2.15

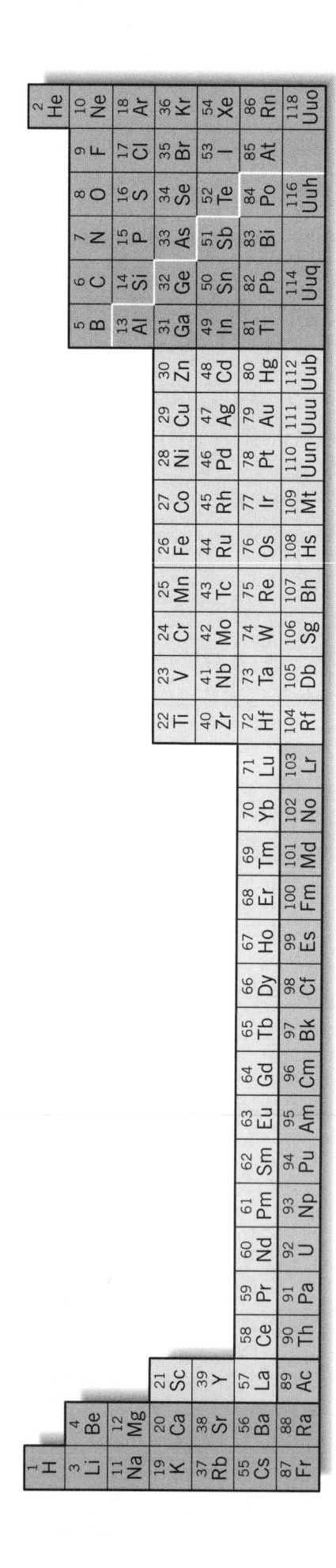

FIGURE 2.16

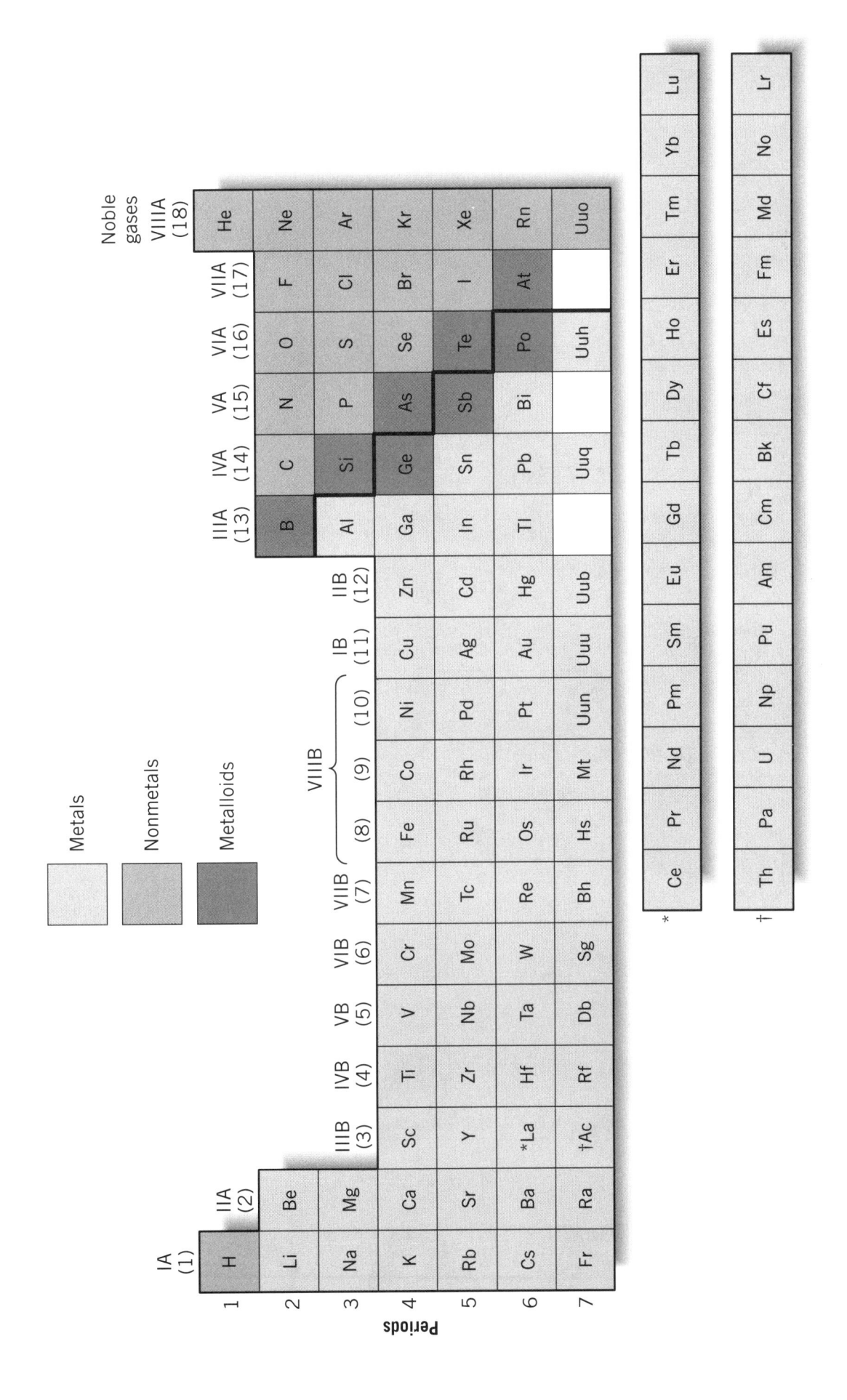

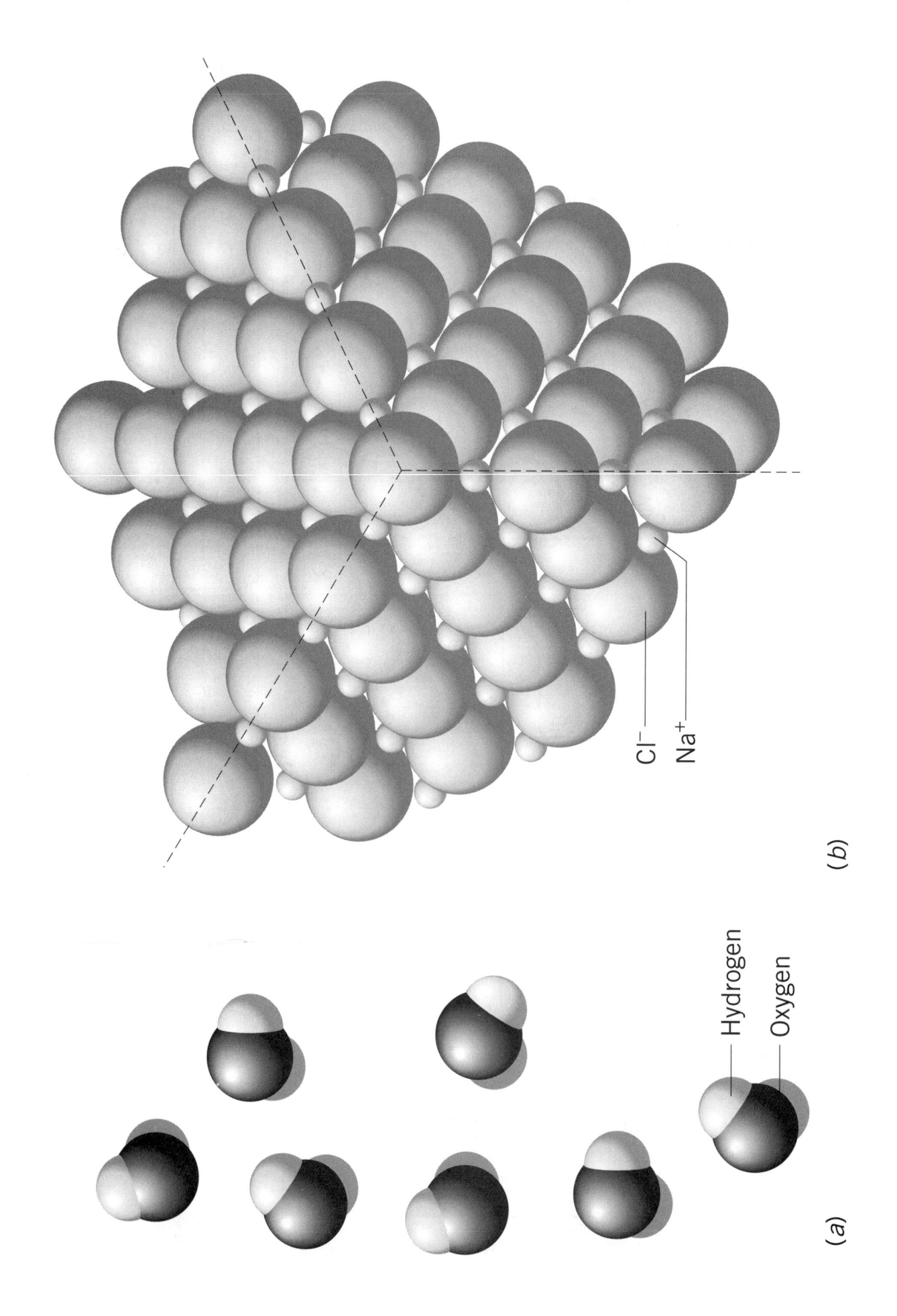

(b)

Cl⁻
Na⁺

Hydrogen
Oxygen

(a)

FIGURE 2.22

propane, C$_3$H$_8$

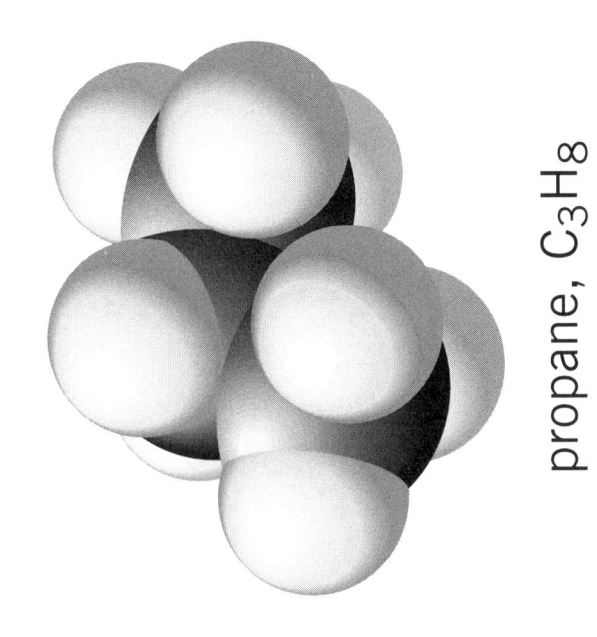

ethane, C$_2$H$_6$

methane, CH$_4$

FIGURE 2.25

FIGURE 4.2

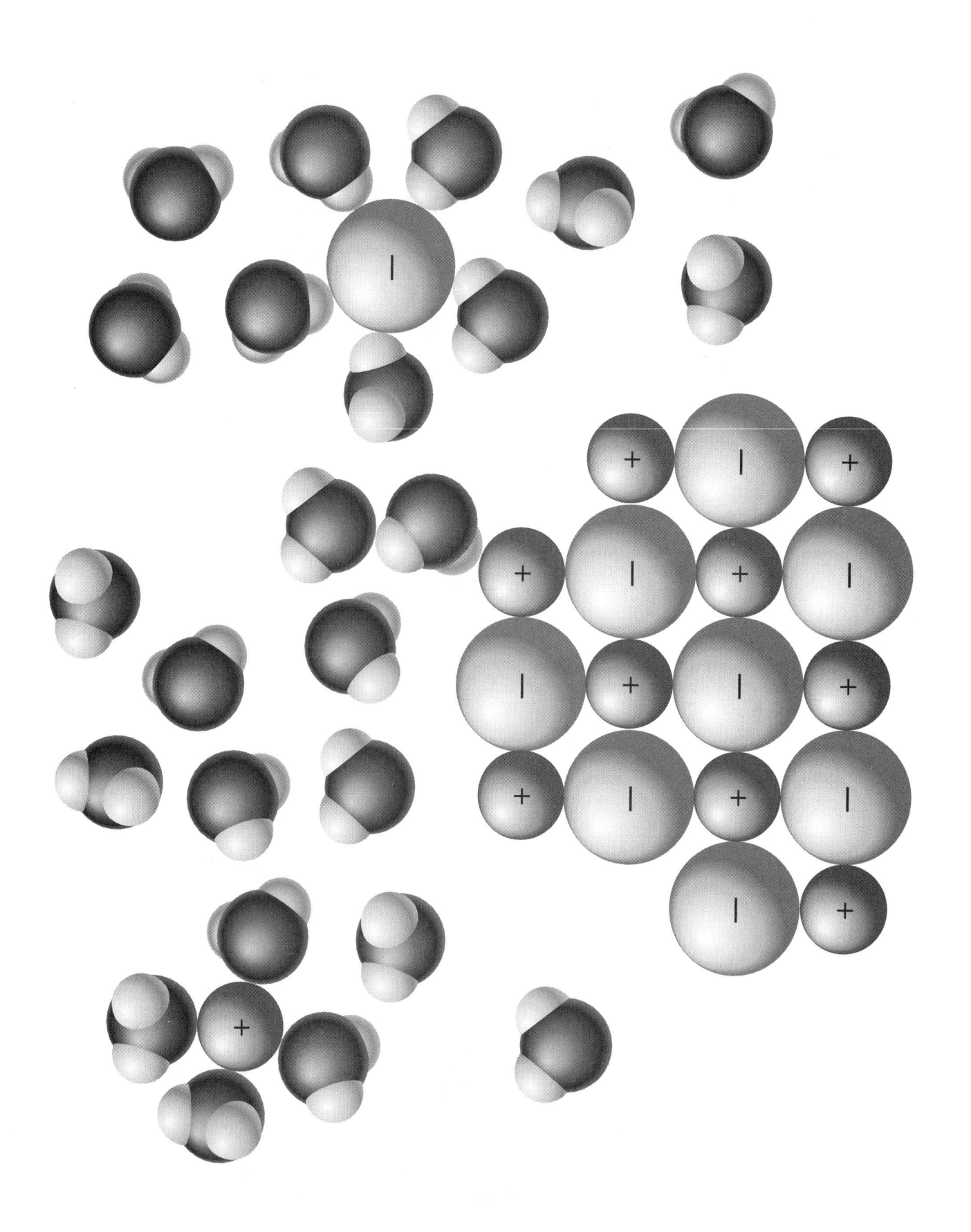

FIGURE 4.3

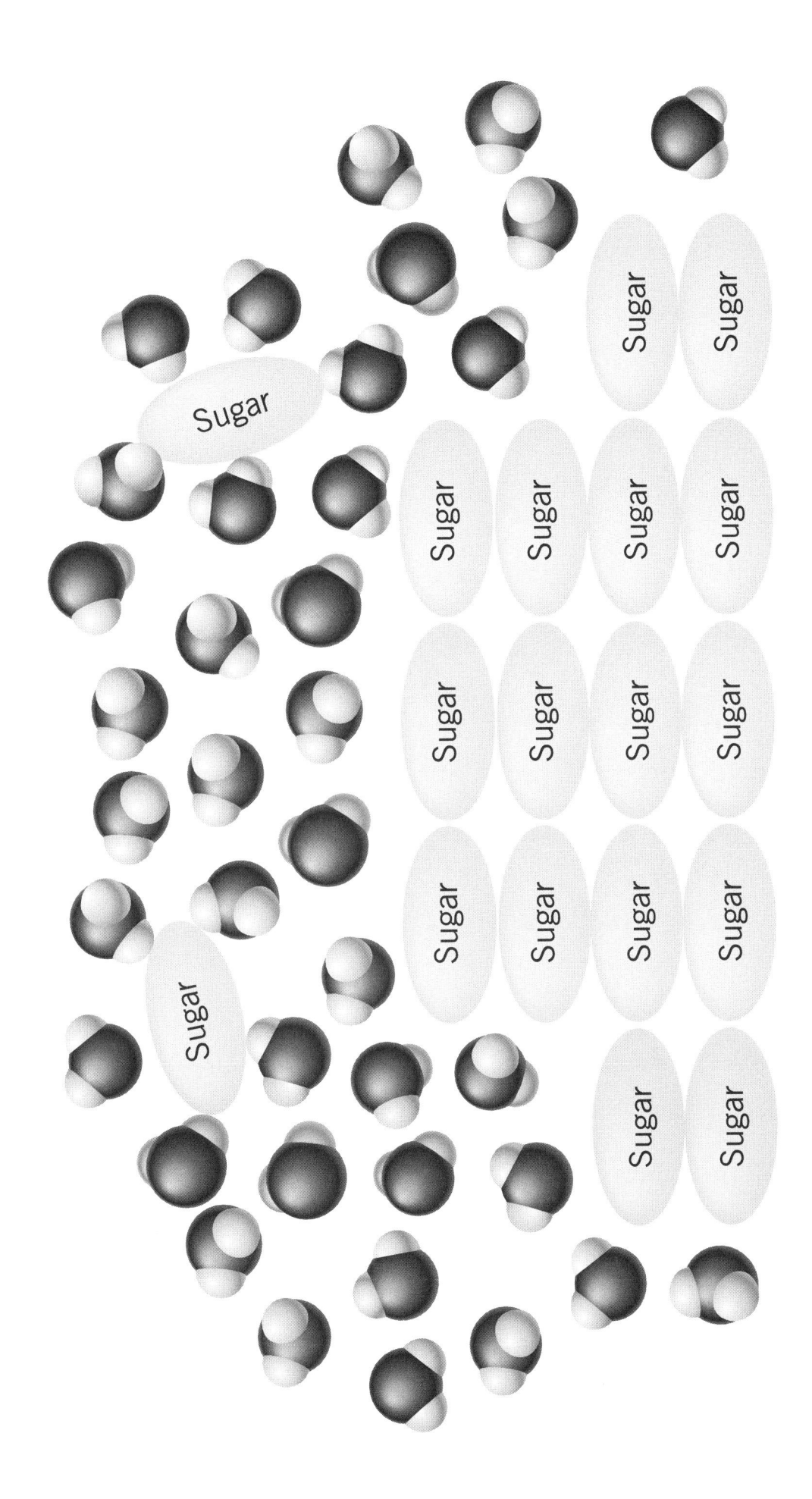

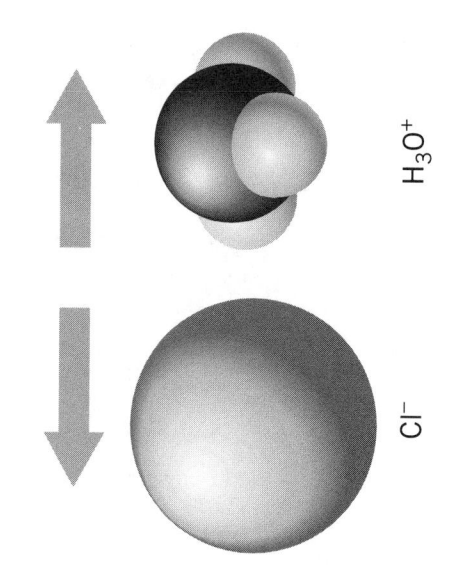

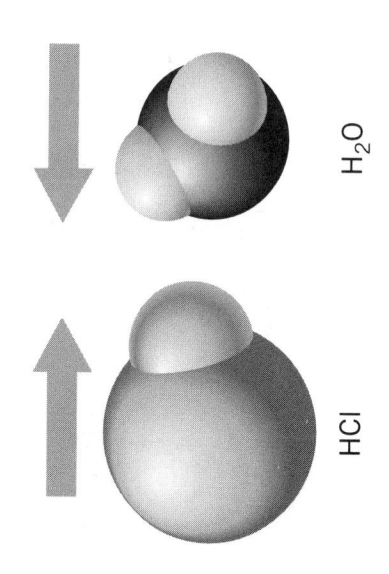

H₃O⁺

Cl⁻

H₂O

HCl

From page 163

OH⁻

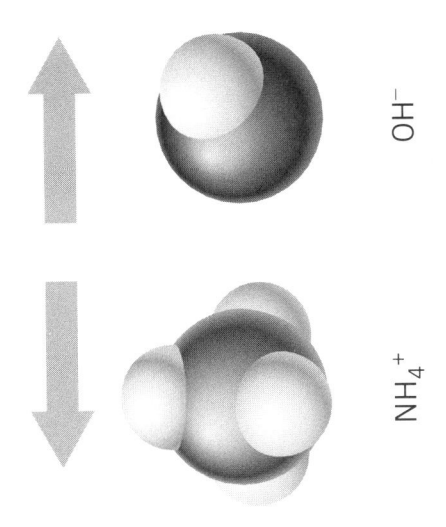

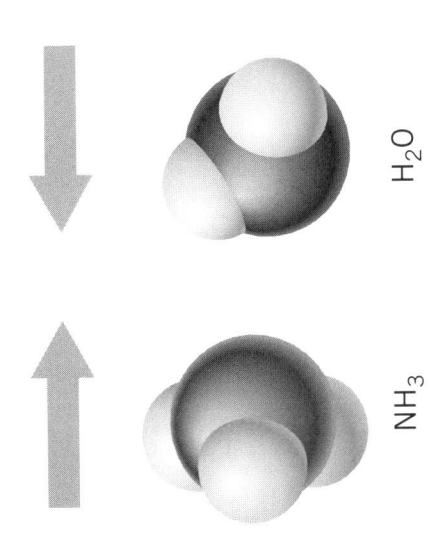

NH₄⁺

H₂O

NH₃

FIGURE 6.1

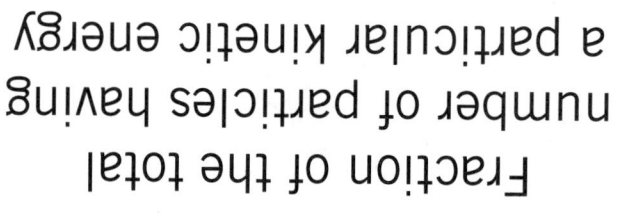

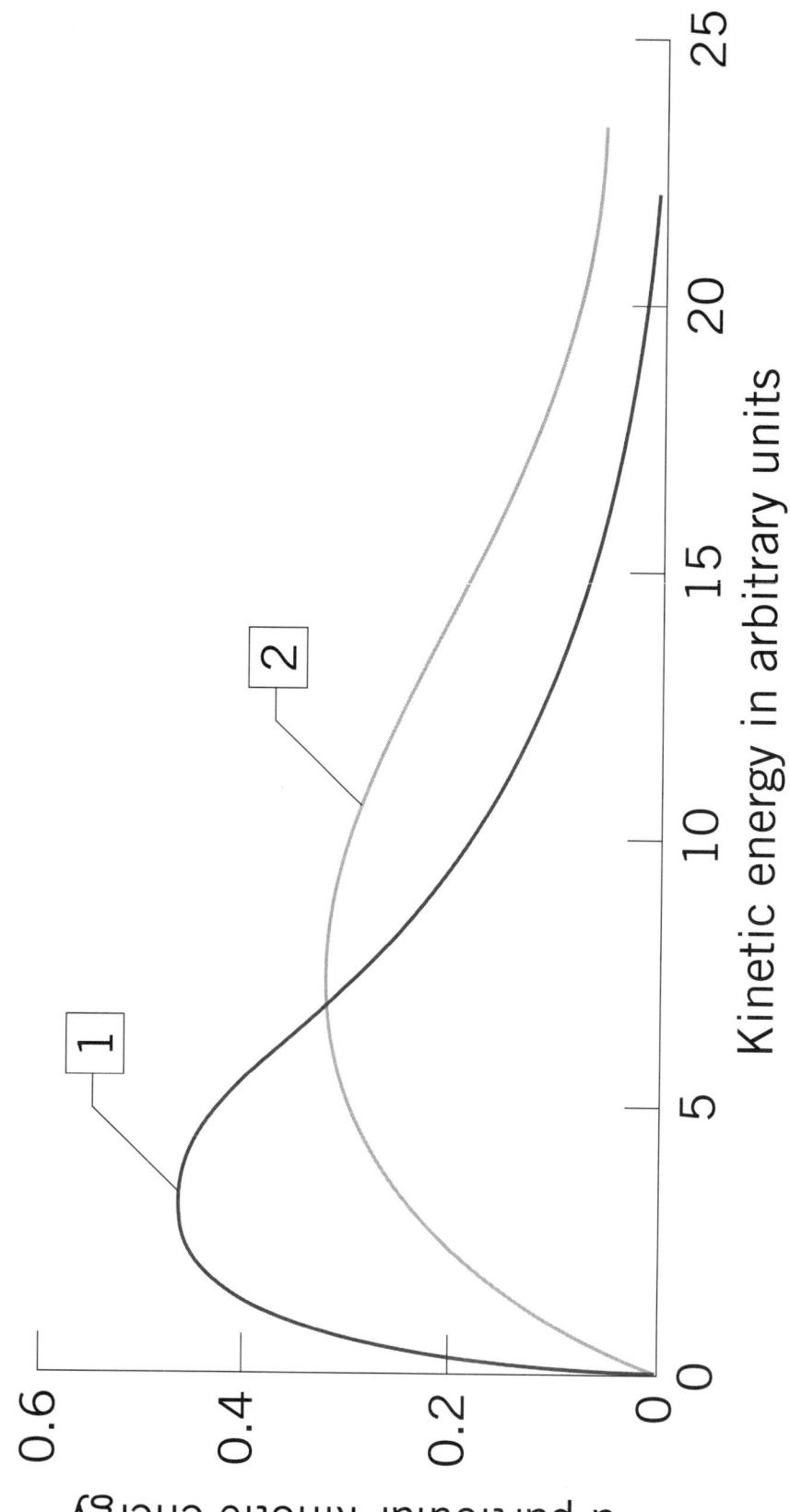

FIGURE 6.3

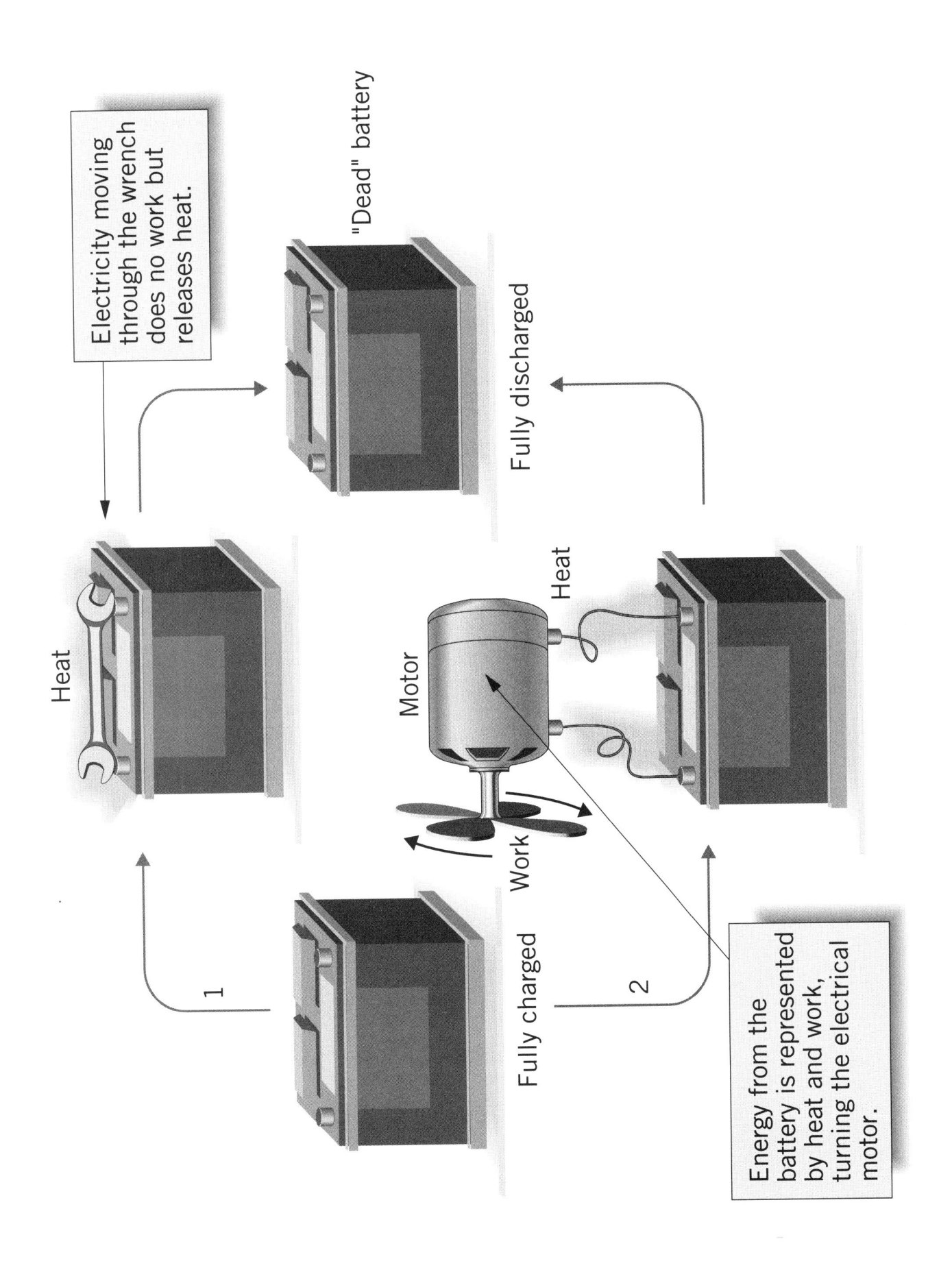

FIGURE 6.4

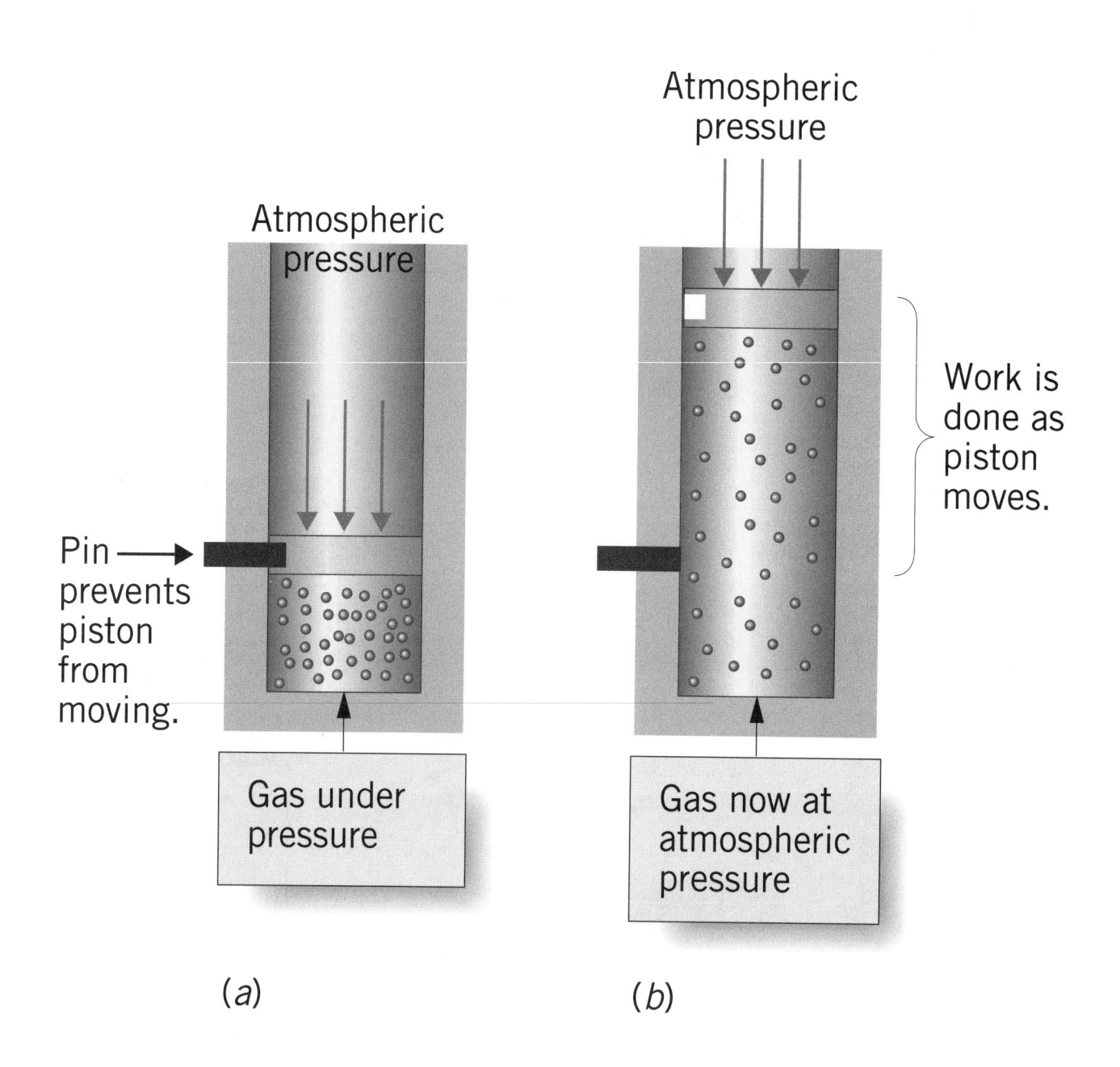

FIGURE 6.5

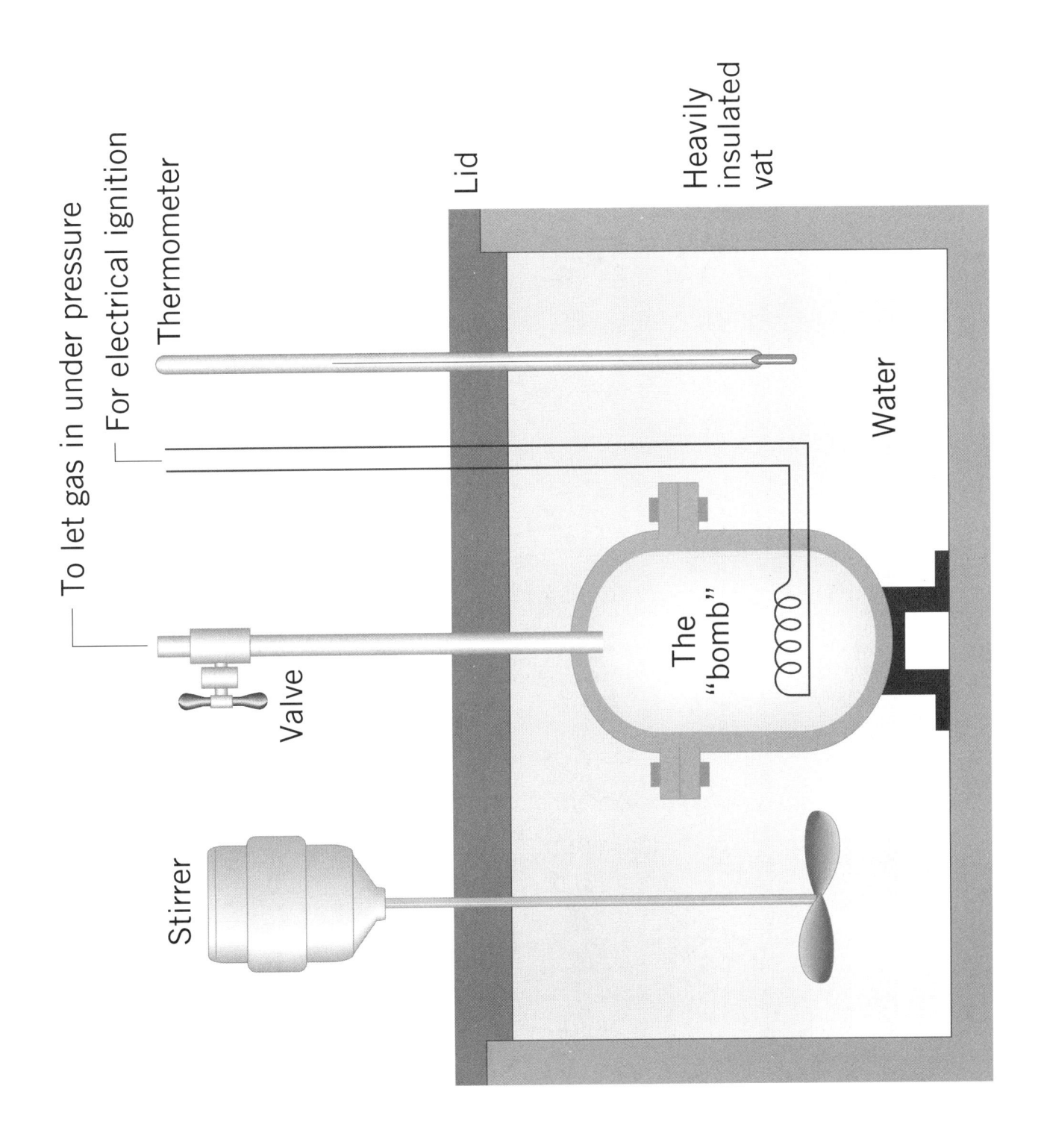

FIGURE 7.3a

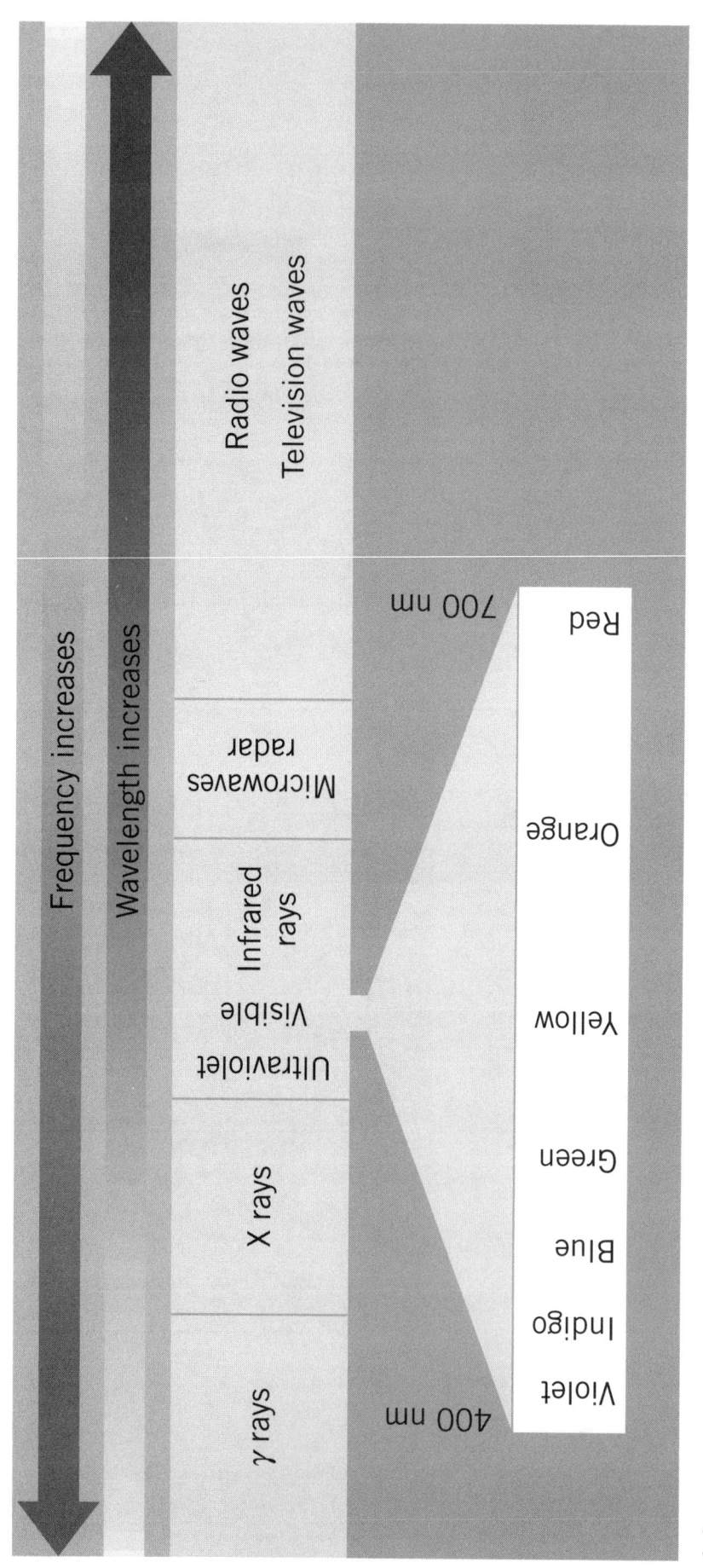

FIGURE 7.6

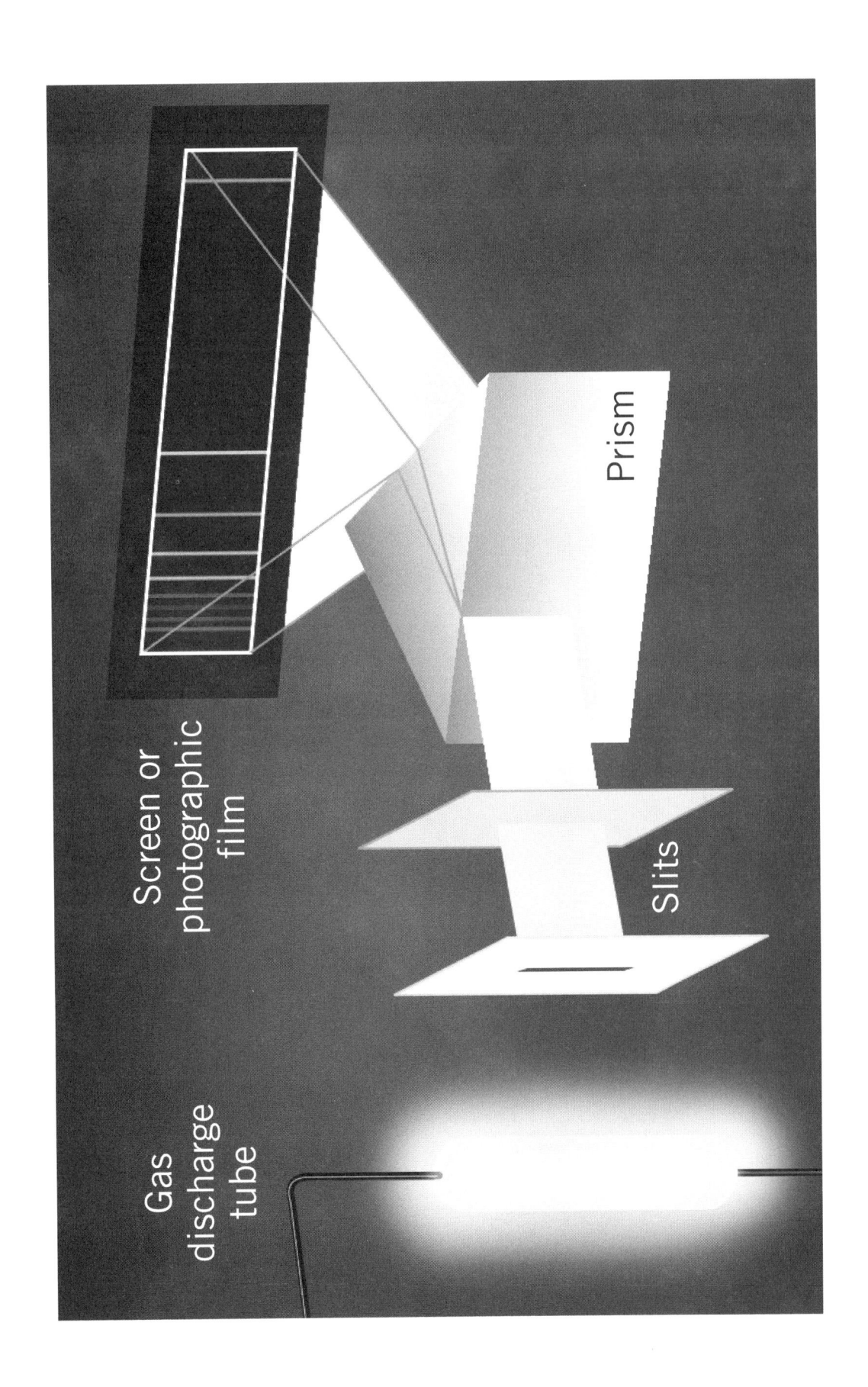

FIGURE 7.15

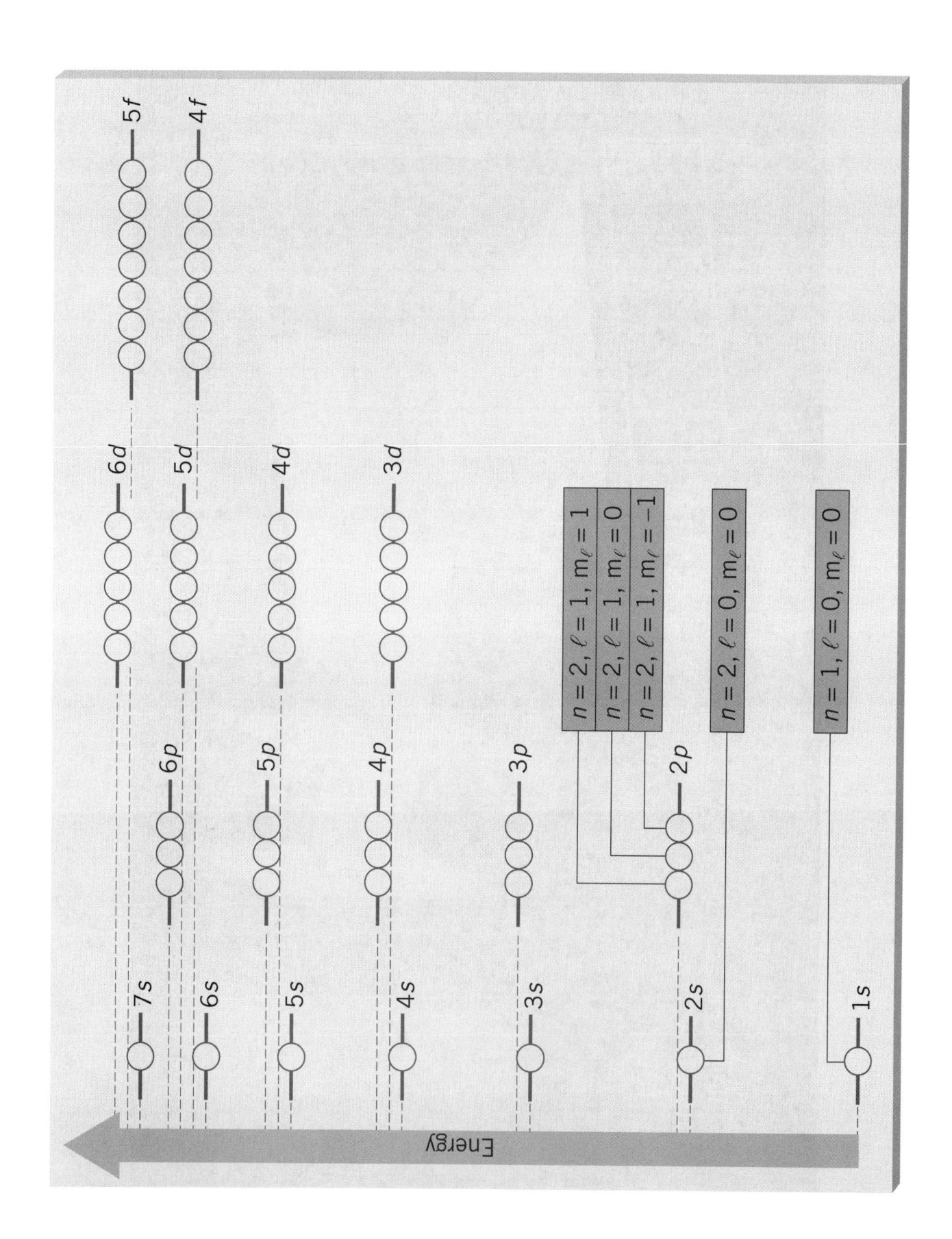

FIGURE 7.17

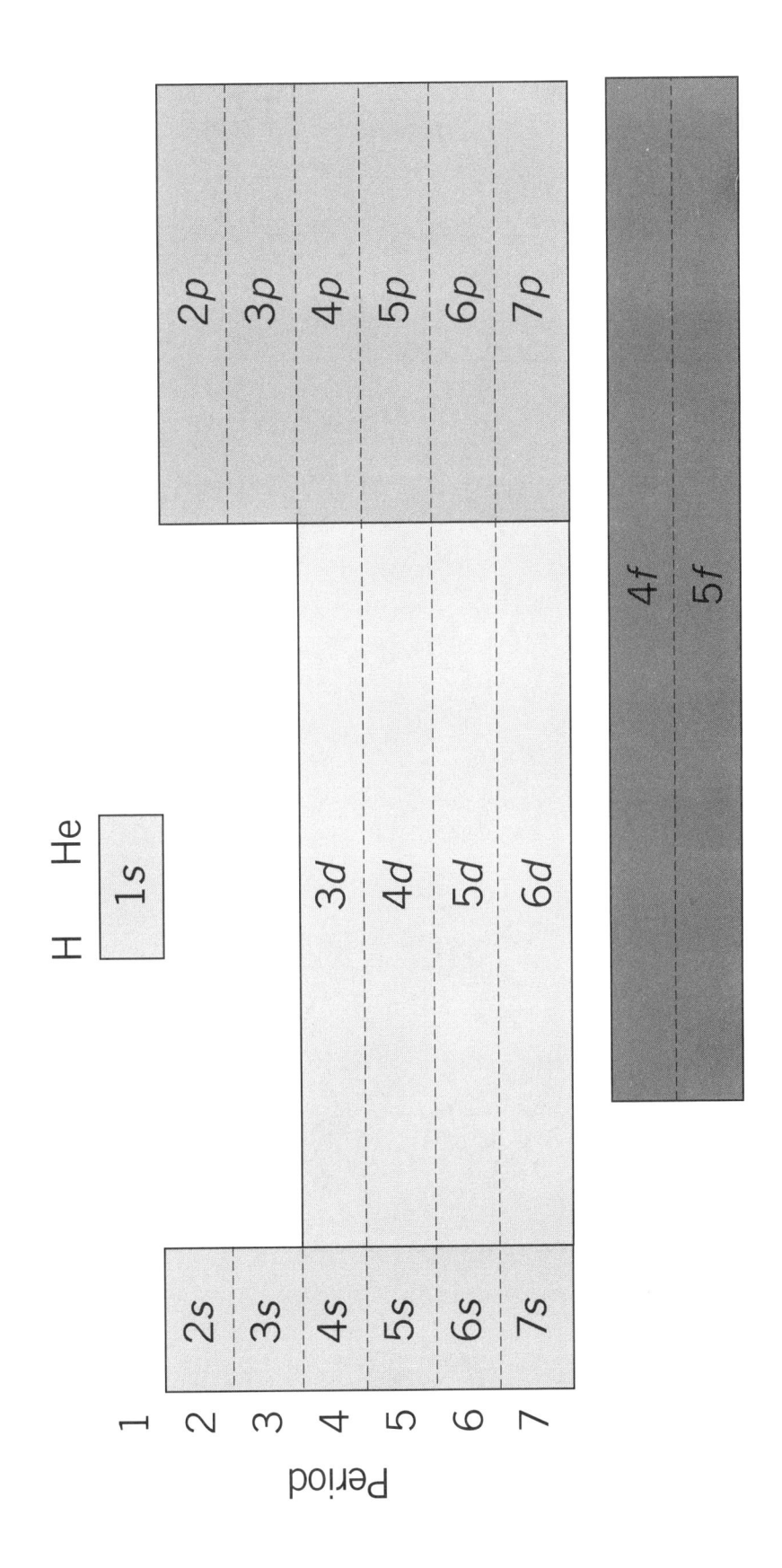

FIGURE 7.19

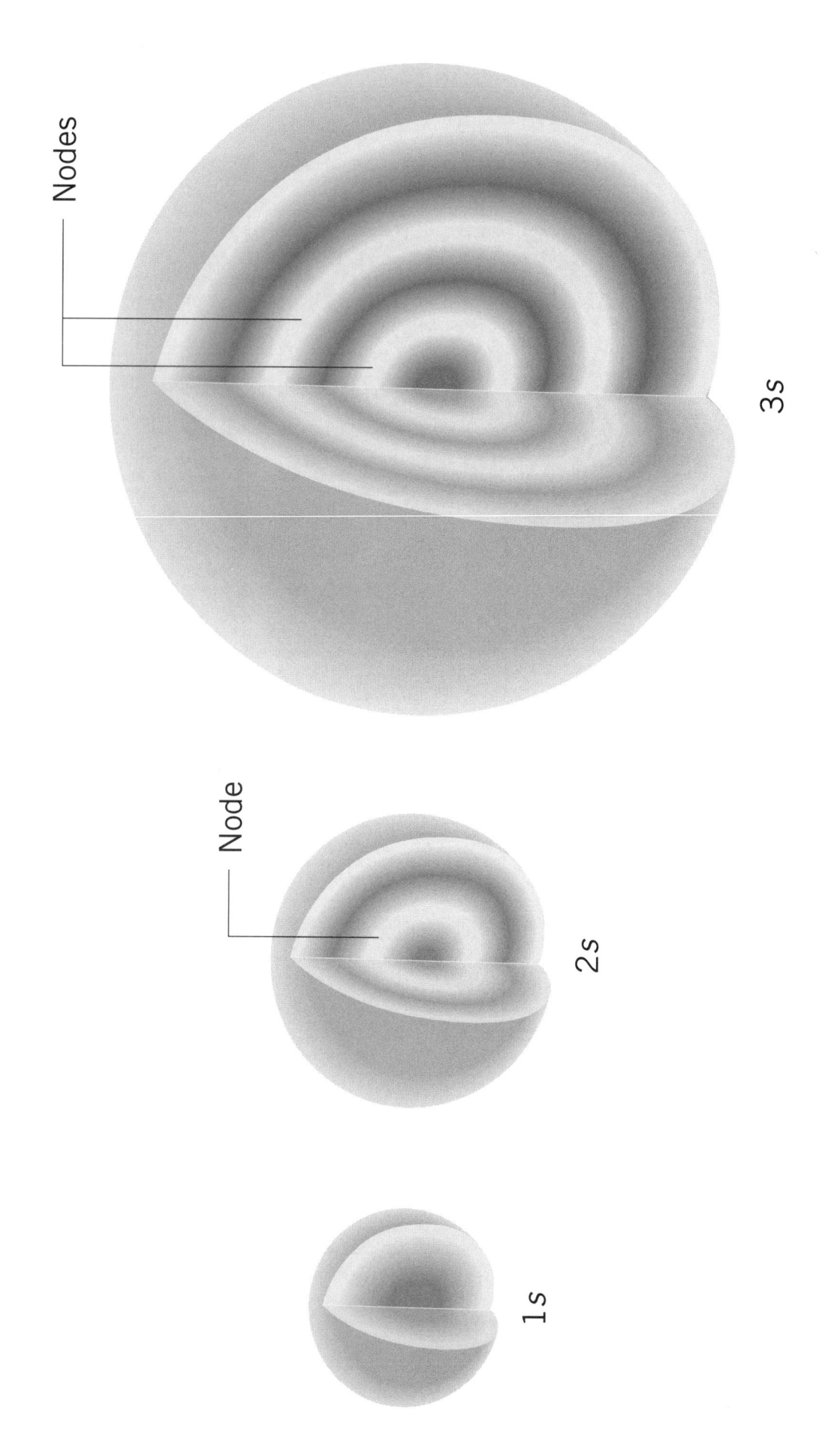

Nodes

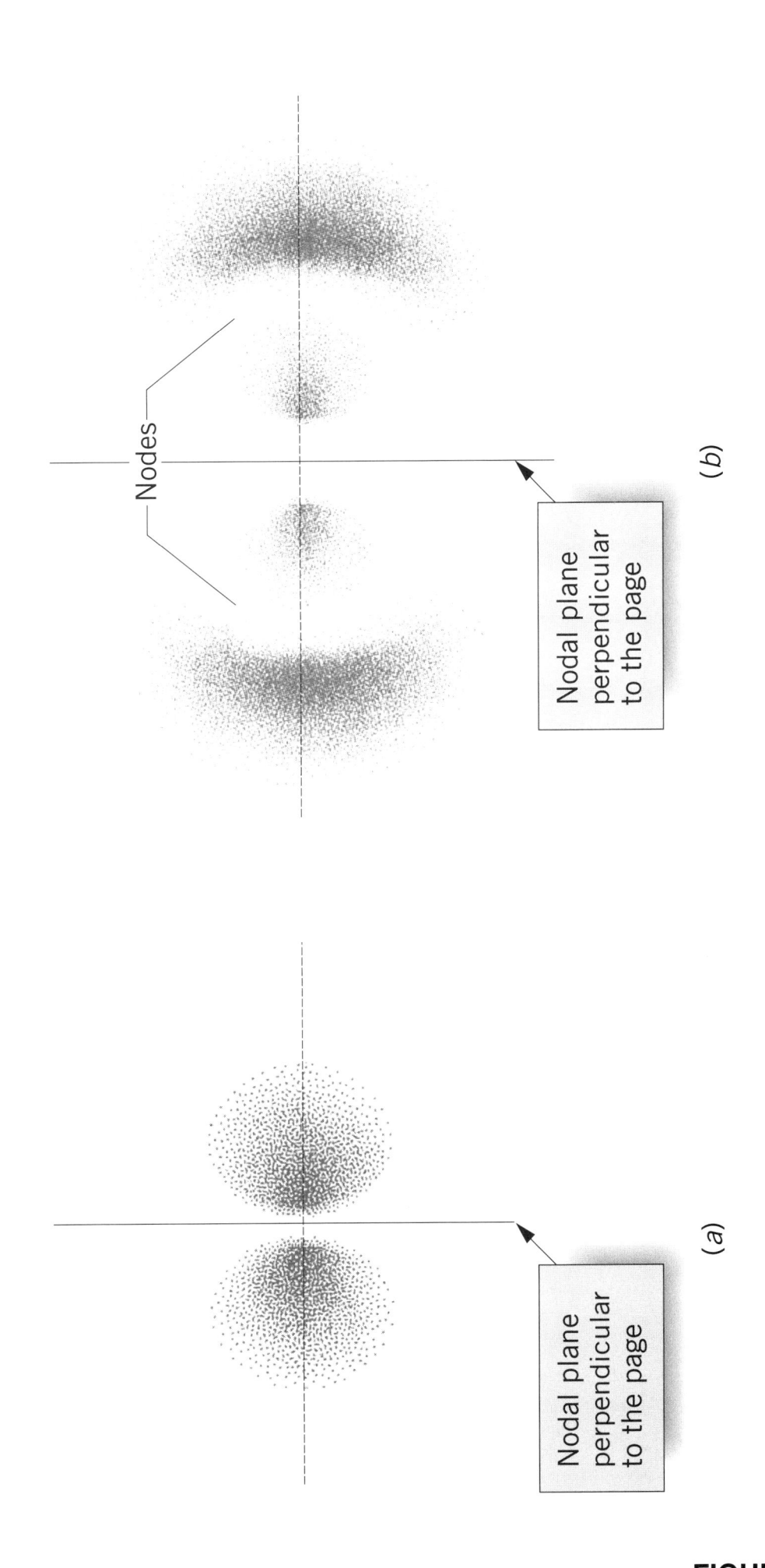

Nodal plane perpendicular to the page

(b)

Nodal plane perpendicular to the page

(a)

FIGURE 7.20

FIGURE 7.22

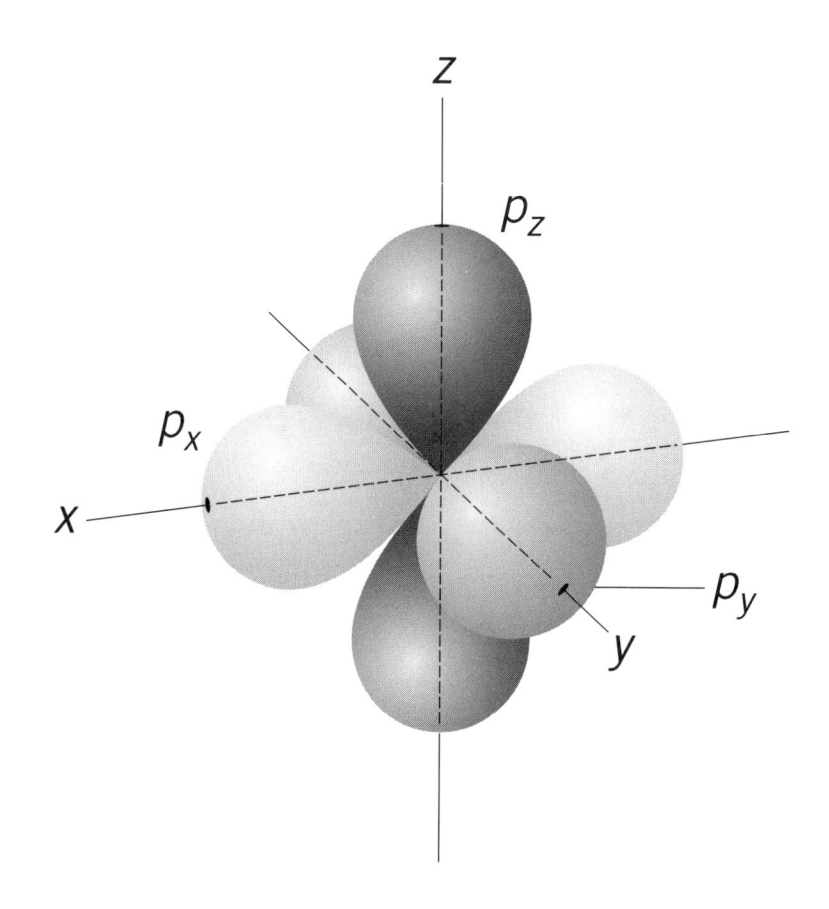

FIGURE 7.23

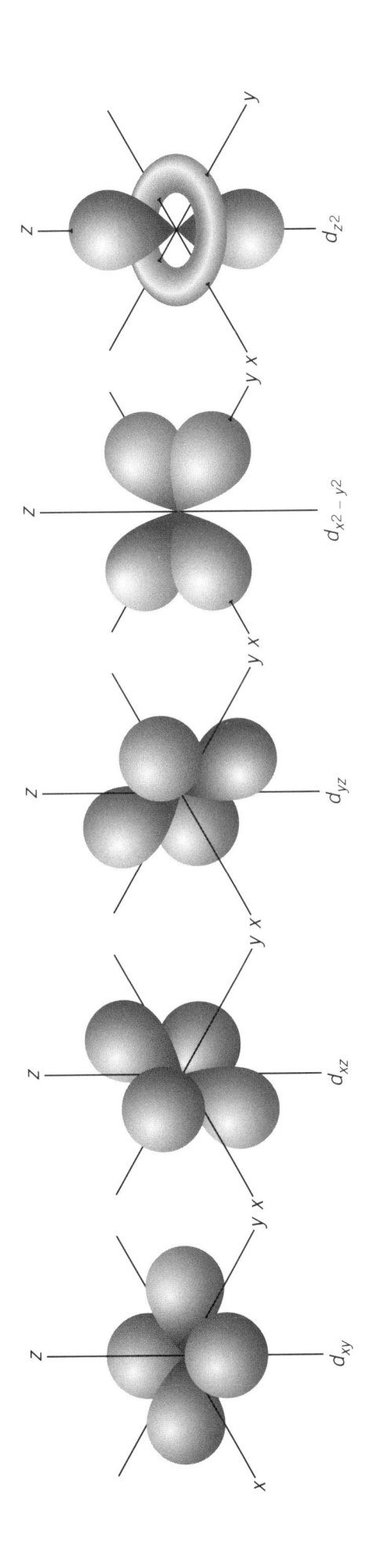

FIGURE 7.25

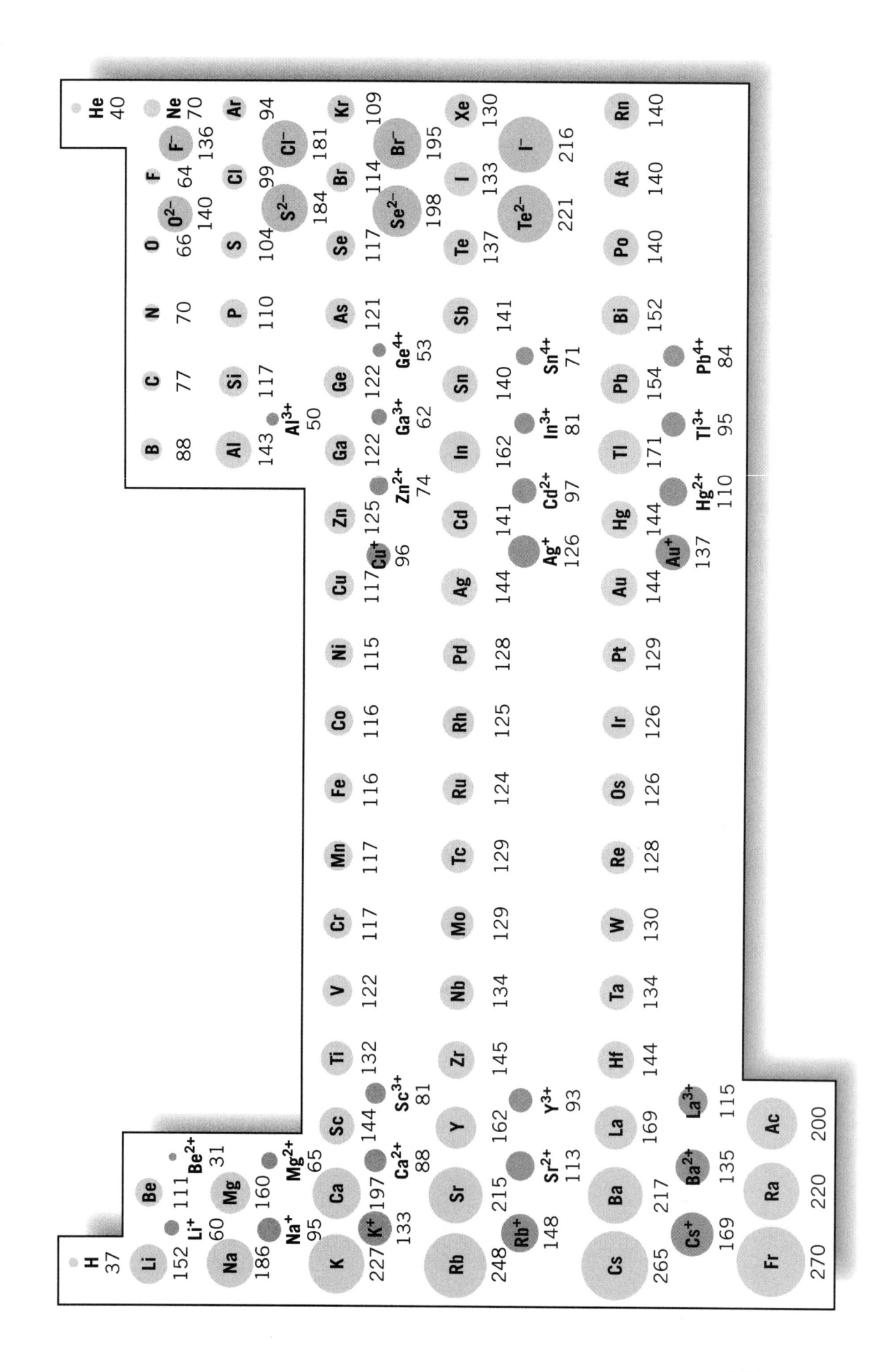

FIGURE 7.27

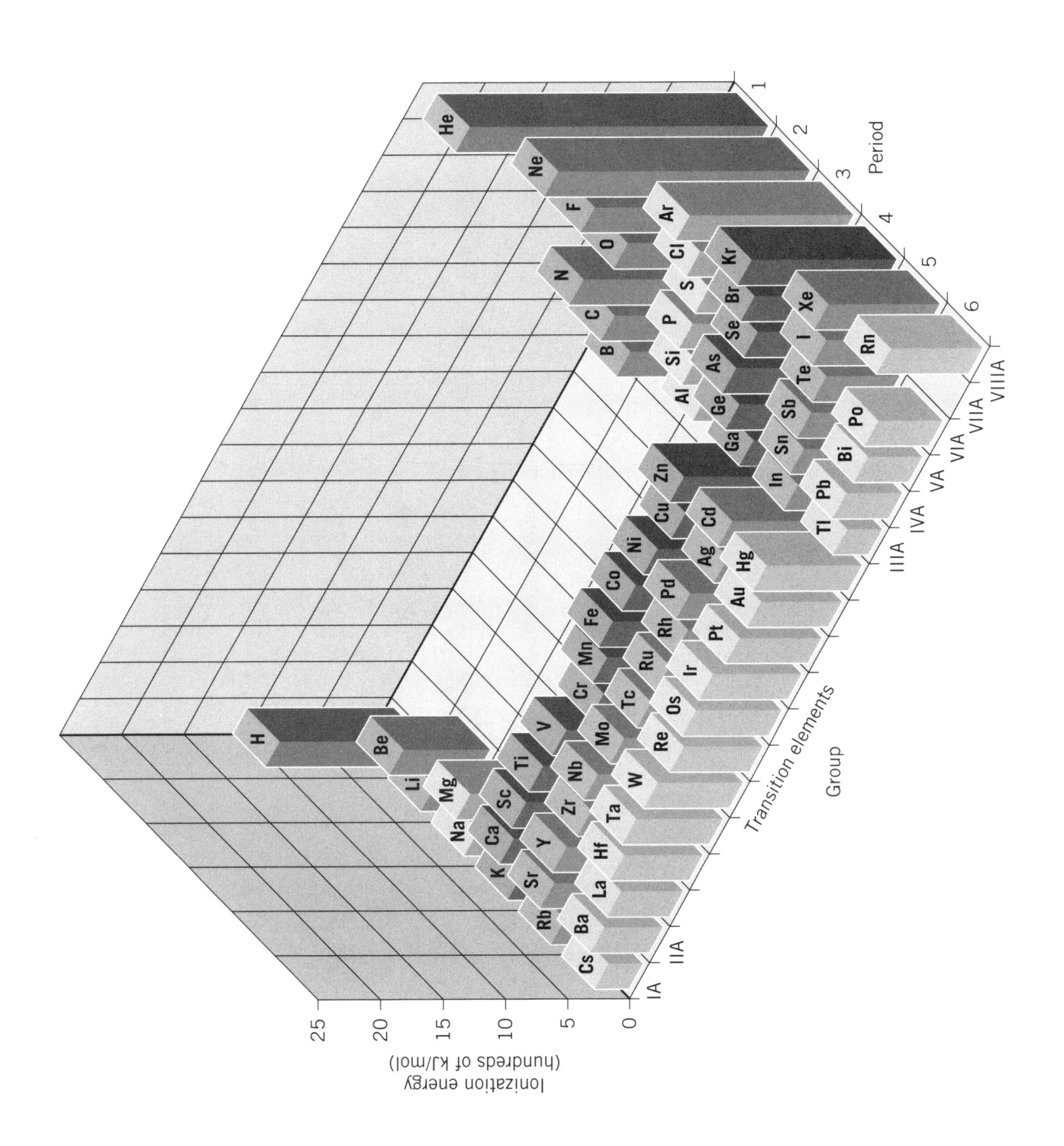

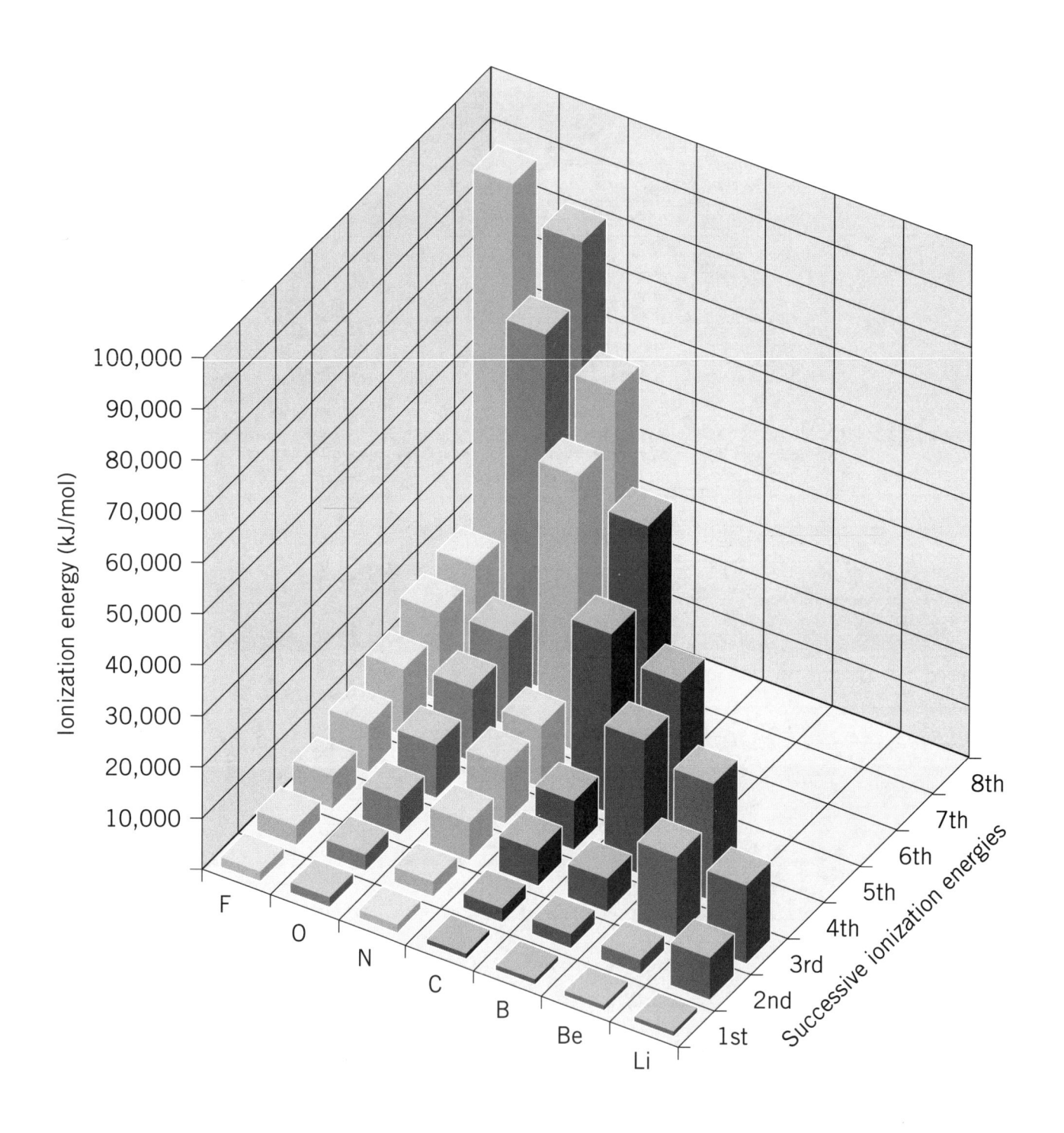

FIGURE 7.28

Enthalpy (energy) ↑

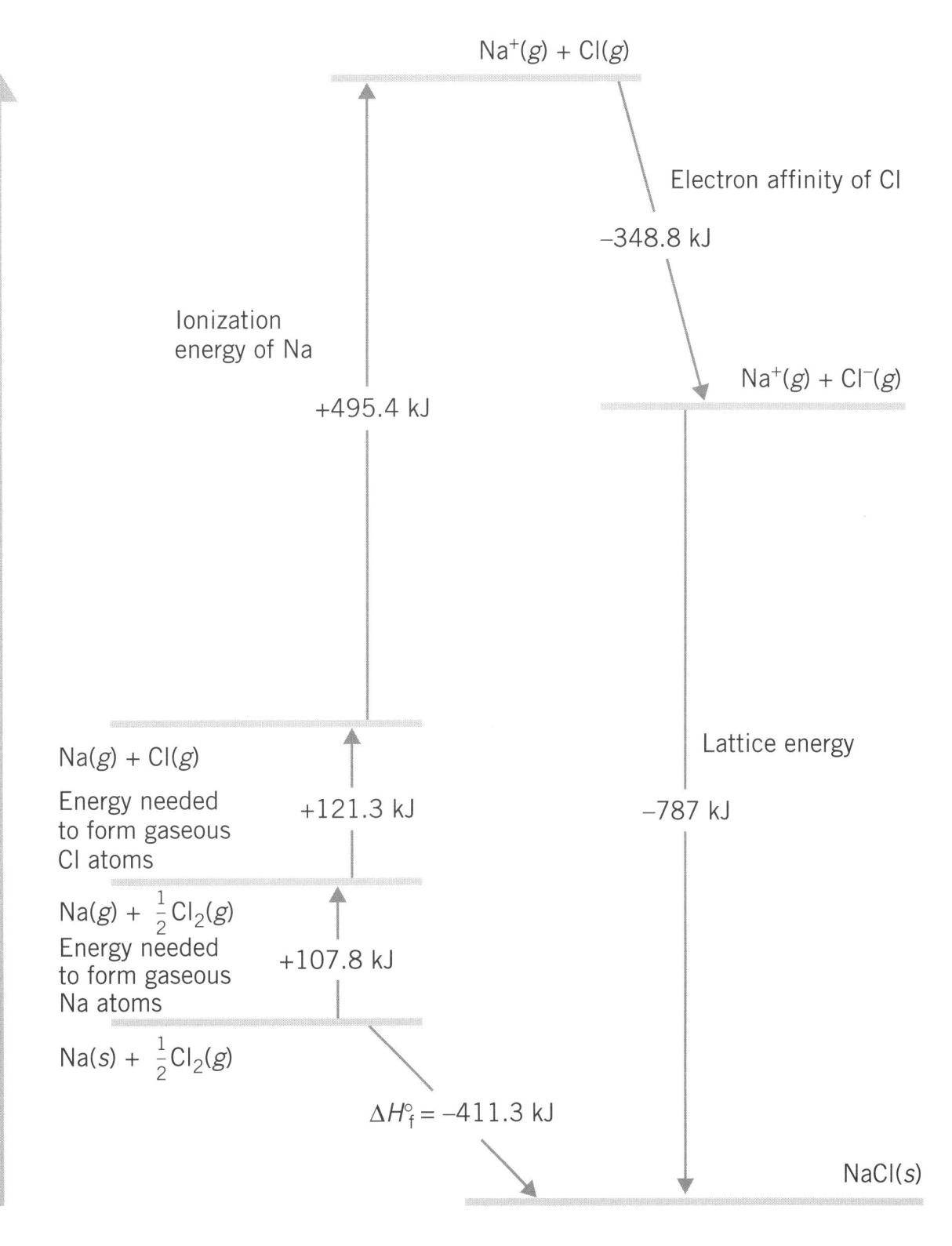

Na⁺(g) + Cl(g)

Electron affinity of Cl
−348.8 kJ

Ionization energy of Na
+495.4 kJ

Na⁺(g) + Cl⁻(g)

Na(g) + Cl(g)
Energy needed to form gaseous Cl atoms
+121.3 kJ

Na(g) + ½Cl₂(g)
Energy needed to form gaseous Na atoms
+107.8 kJ

Lattice energy
−787 kJ

Na(s) + ½Cl₂(g)

$\Delta H°_f = -411.3$ kJ

NaCl(s)

FIGURE 8.1

FIGURE 8.4

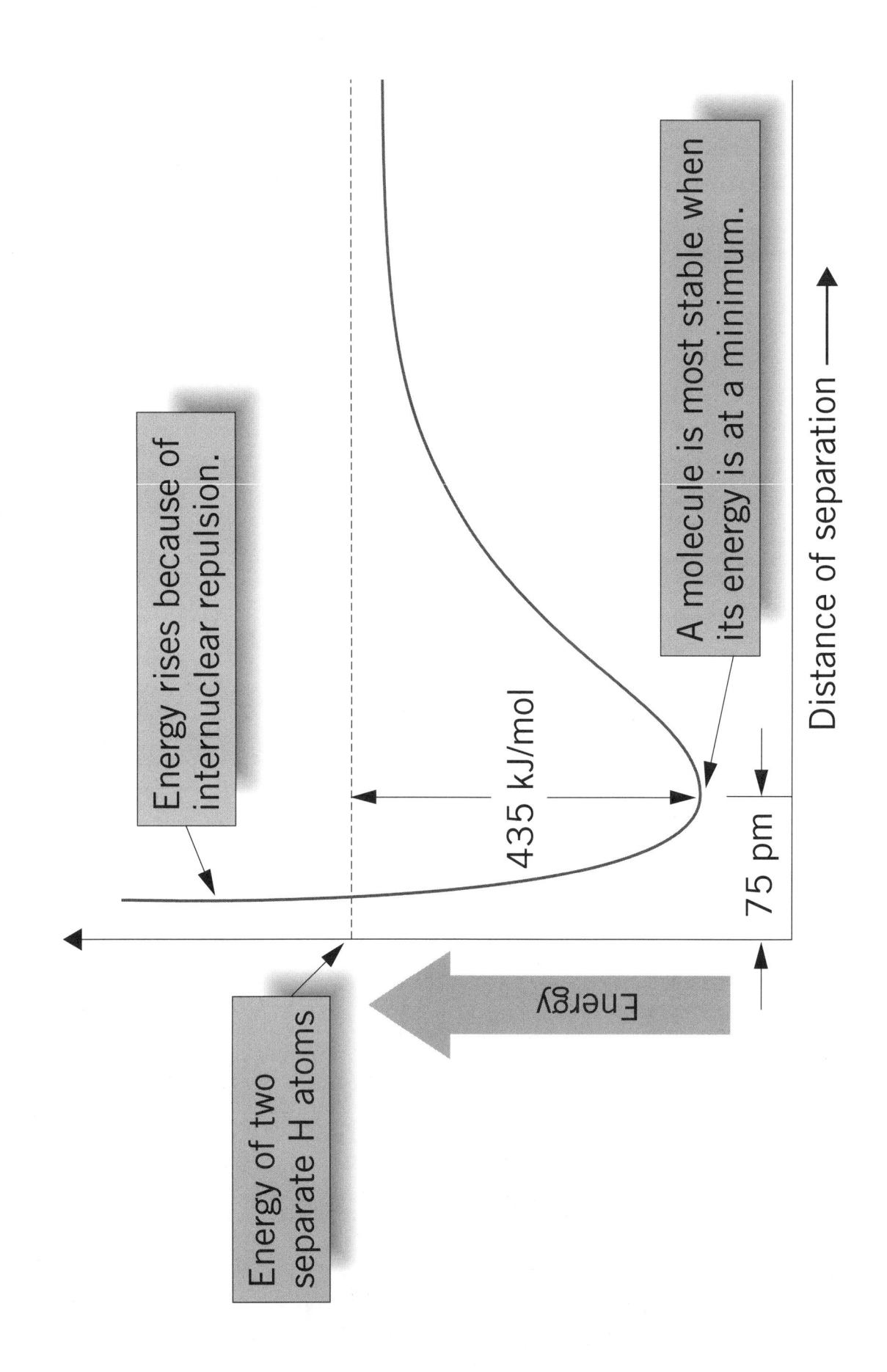

FIGURE 8.6

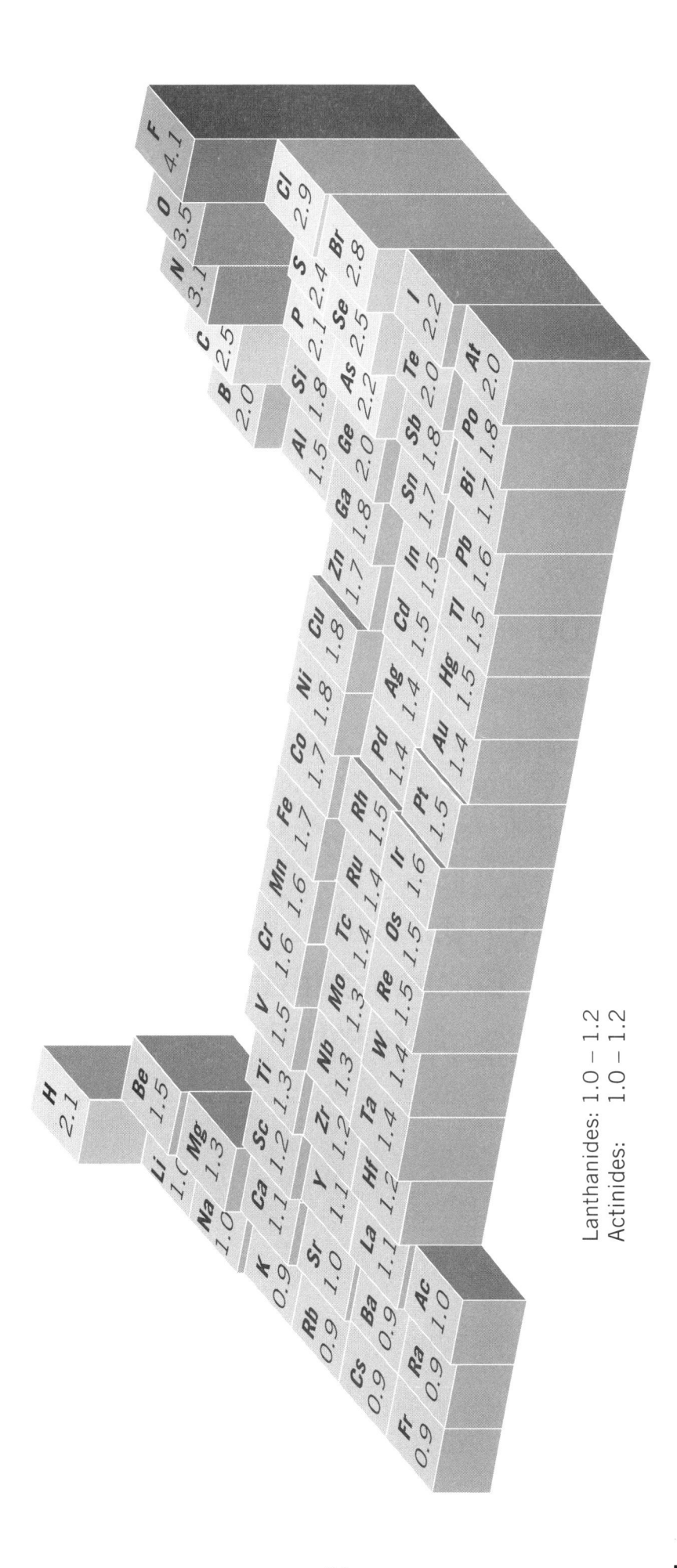

Lanthanides: 1.0 – 1.2
Actinides: 1.0 – 1.2

FIGURE 8.7

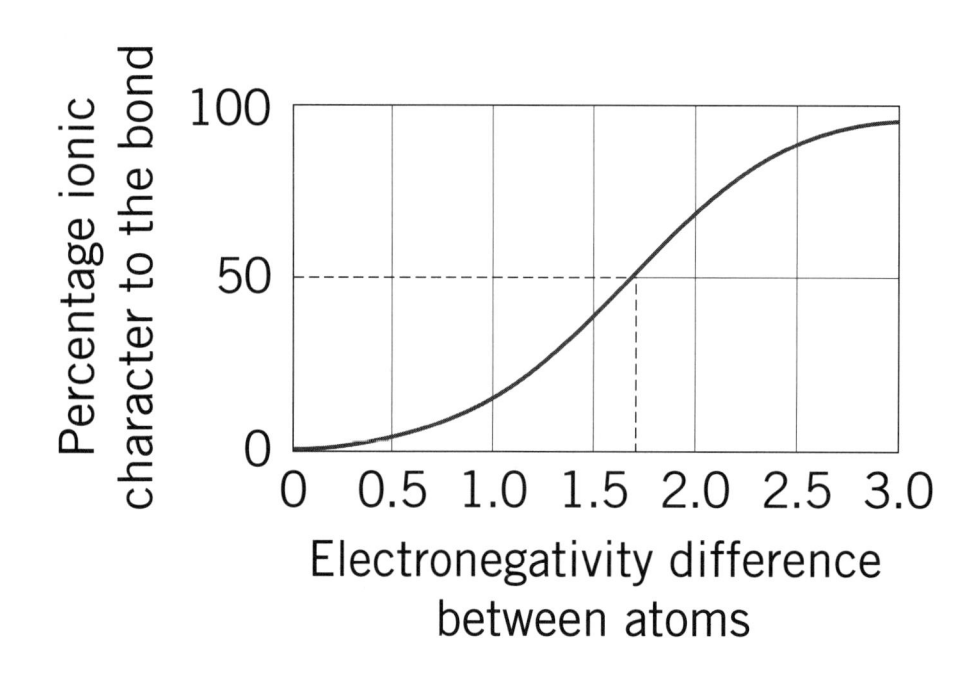

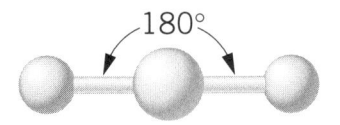

180°
A linear molecule

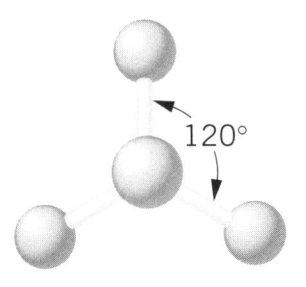

120°
A planar triangular molecule

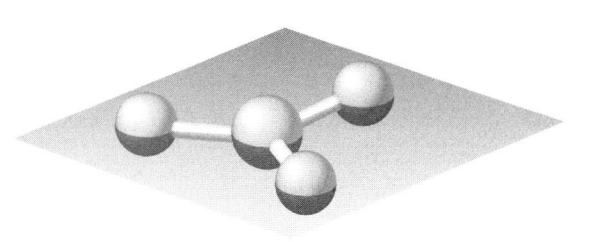

Another view showing how all the atoms are in the same plane

A tetrahedron

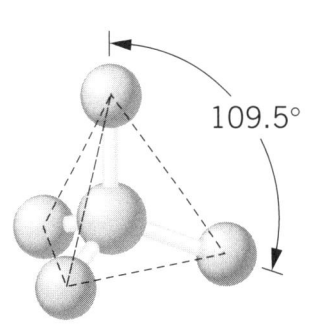

109.5°
A tetrahedral molecule

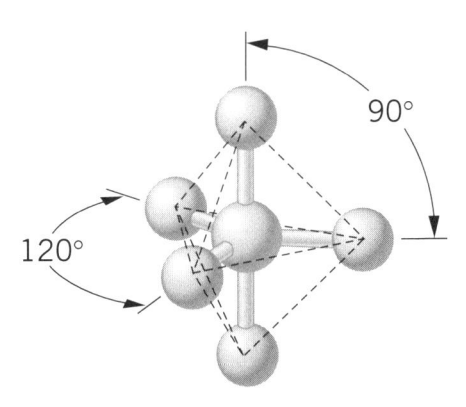

90°
120°
A trigonal bipyramidal molecule

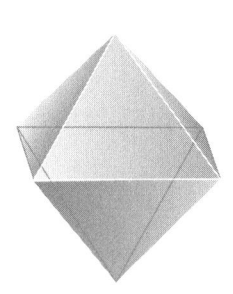

An octahedron

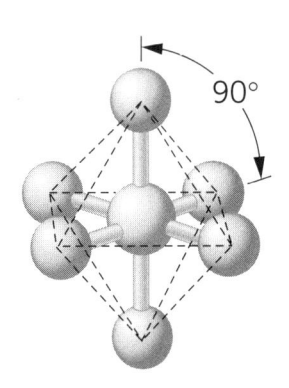

90°
An octahedral molecule

Copyright © 2000 John Wiley & Sons, Inc.

From pages 368, 369 & 370

Number of Electron Pairs	Shape		Example
2	 Linear		BeCl$_2$ — 180° Cl–Be–Cl 
3	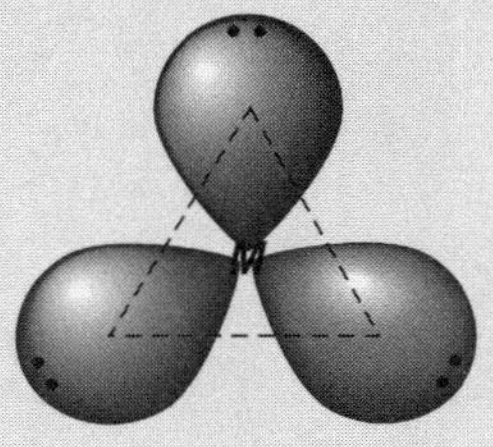 Planar triangular		BCl$_3$ — 120° 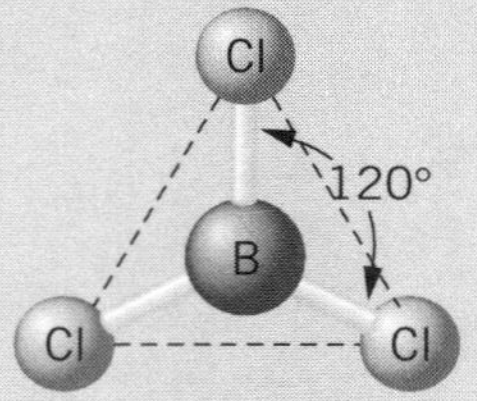
4	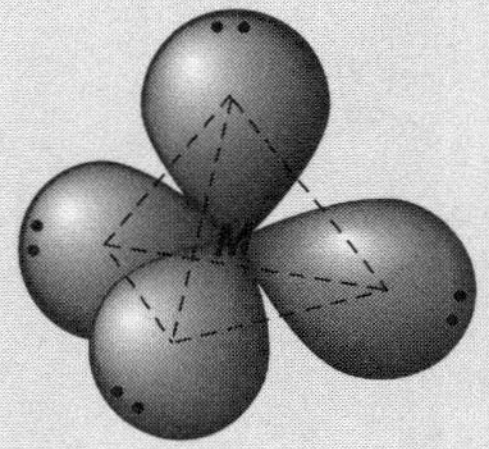 Tetrahedral (A tetrahedron is pyramid shaped. It has four triangular faces and four corners.)		CH$_4$ — 109.5° 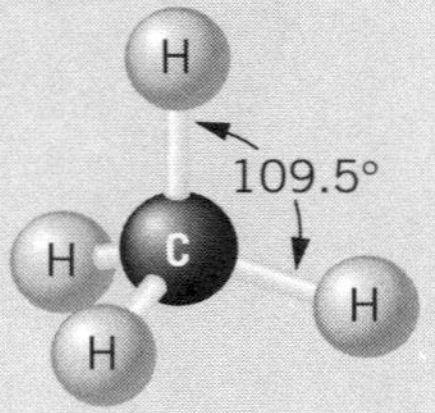
5	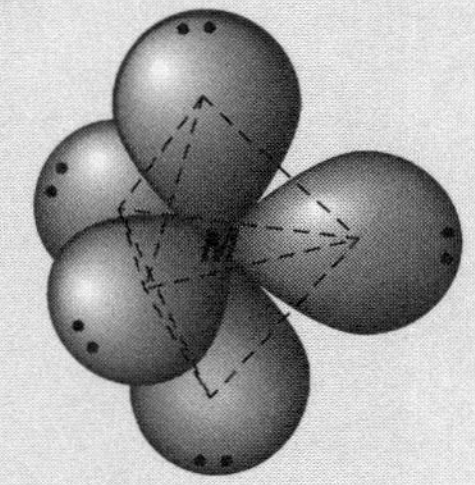 Trigonal bipyramidal (This figure consists of two three-sided pyramids joined by sharing a common face—the triangular plane through the center.)		PCl$_5$ 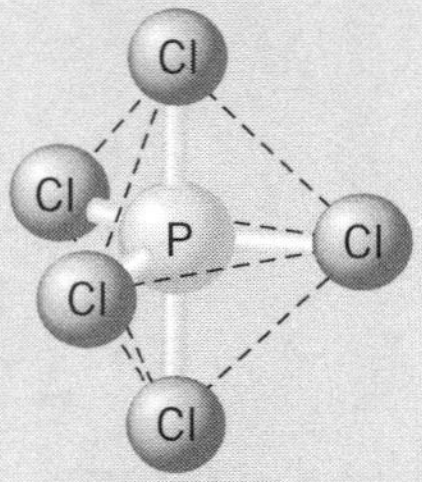
6	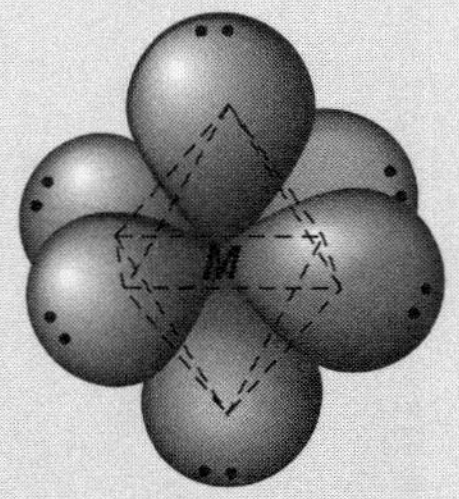 Octahedral (An octahedron is an eight-sided figure with *six* corners. It consist of two square pyramids that share a common square base.)		SF$_6$ 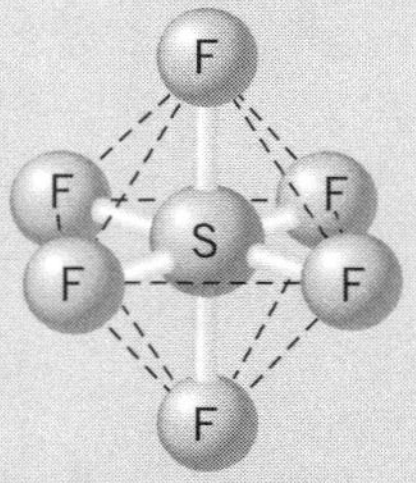

FIGURE 9.2

Number of Pairs in Bonds	Number of Lone Pairs	Structure
4	0	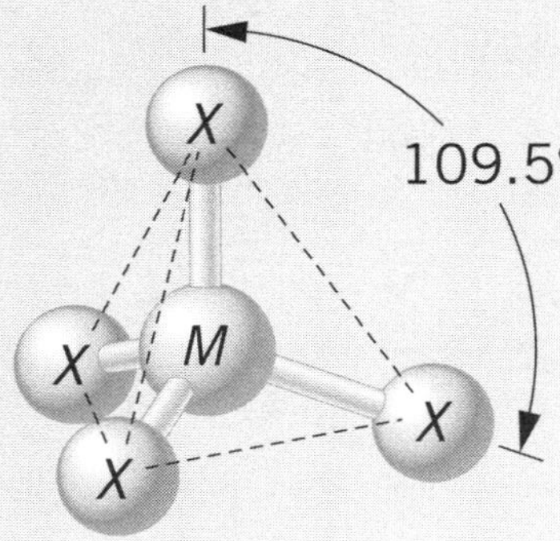 **Tetrahedral** (example, CH_4) All bond angles are 109.5°.
3	1	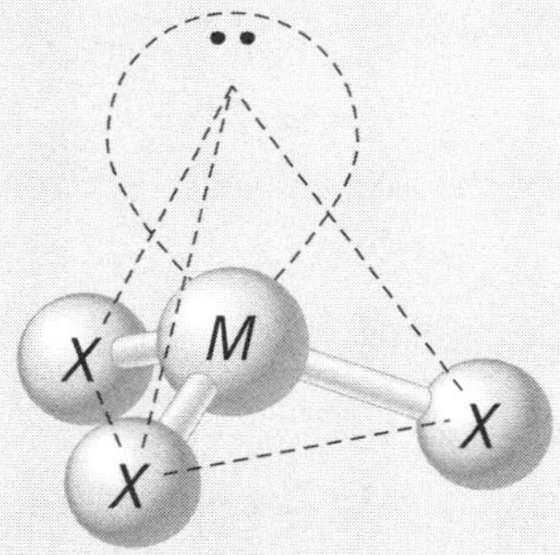 **Trigonal pyramidal** (pyramid shaped) (example, NH_3)
2	2	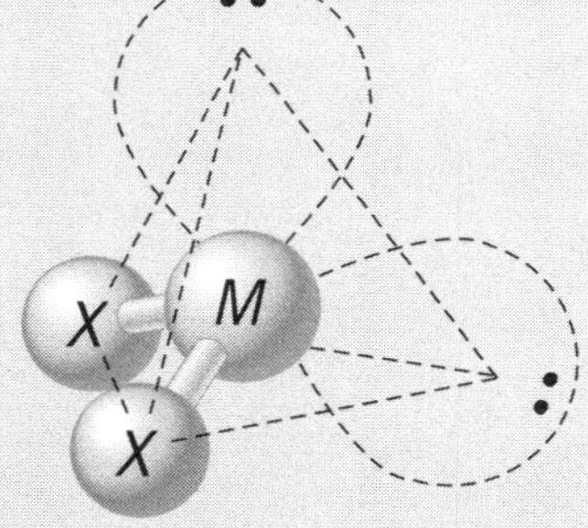 **Nonlinear, bent** (example, H_2O)

FIGURE 9.3

Number of Pairs in Bonds	Number of Lone Pairs	Structure	
5	0	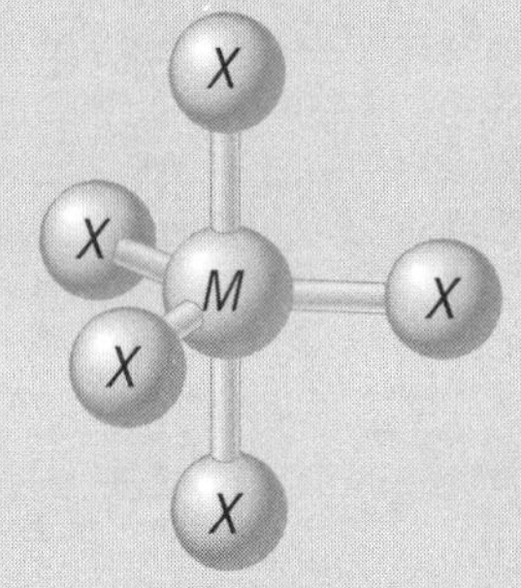	**Trigonal bipyramidal** (example, PCl$_5$)
4	1	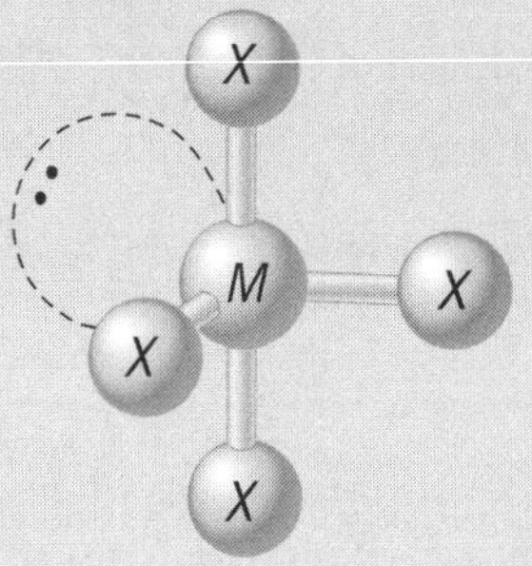	**Distorted tetrahedral** 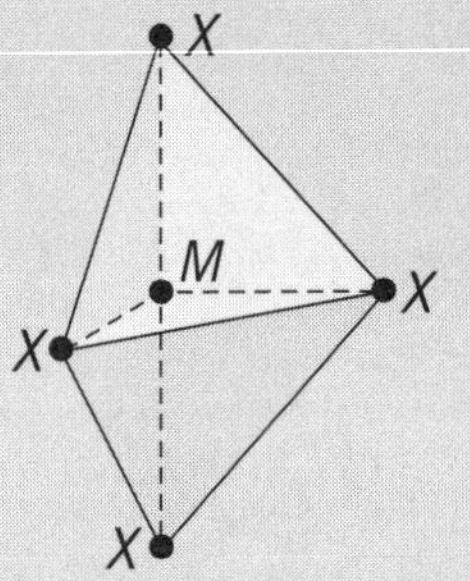 (example, SF$_4$)
3	2	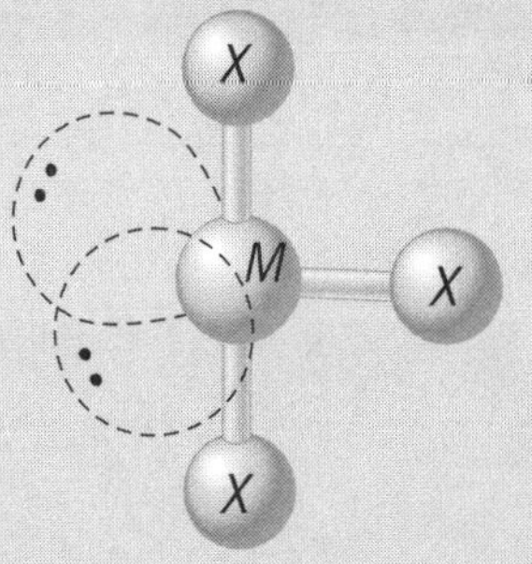	**T-shaped** 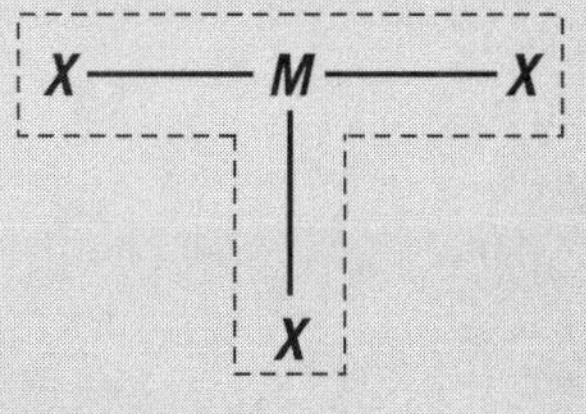 (example, ClF$_3$)
2	3	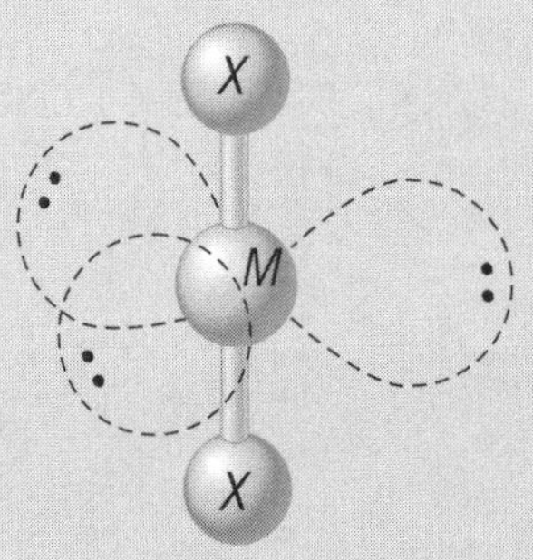	**Linear** (example, I$_3^-$)

FIGURE 9.4

Number of Pairs in Bonds	Number of Lone Pairs	Structure	
6	0	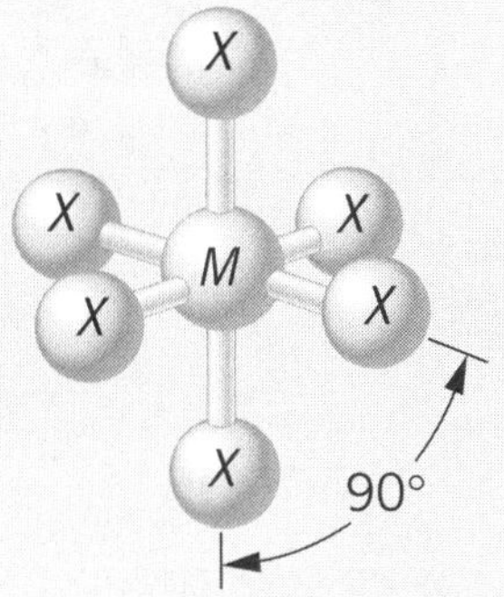	**Octahedral** (example, SF_6) all bond angles are 90°
5	1	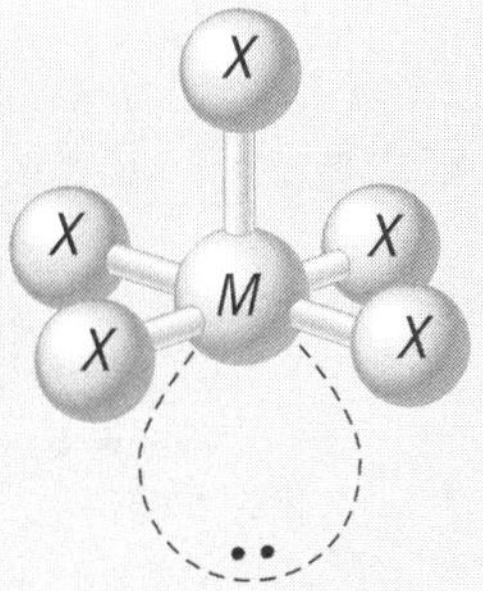	**Square pyramidal** 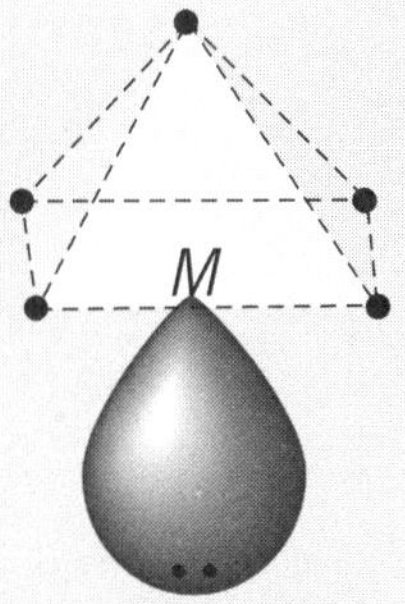 (example, BrF_5)
4	2	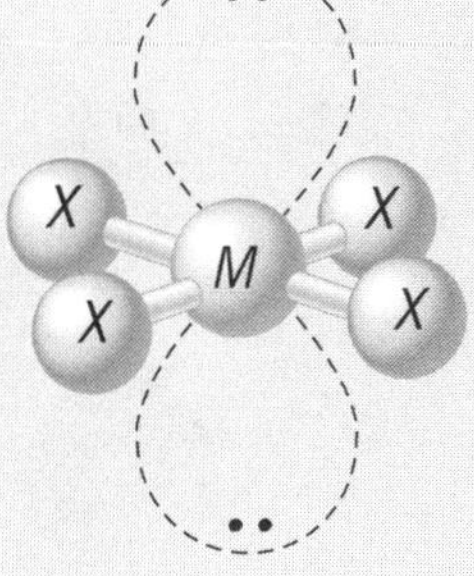	**Square planar** (example, XeF_4)

FIGURE 9.5

CCl₄

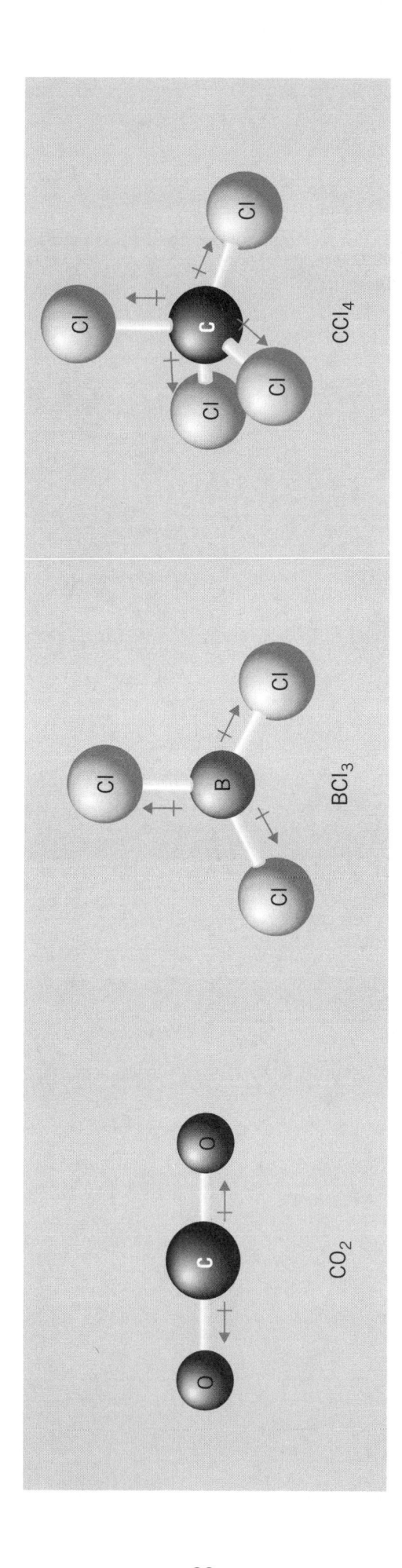

BCl₃

CO₂

Copyright © 2000 John Wiley & Sons, Inc.

FIGURE 9.7

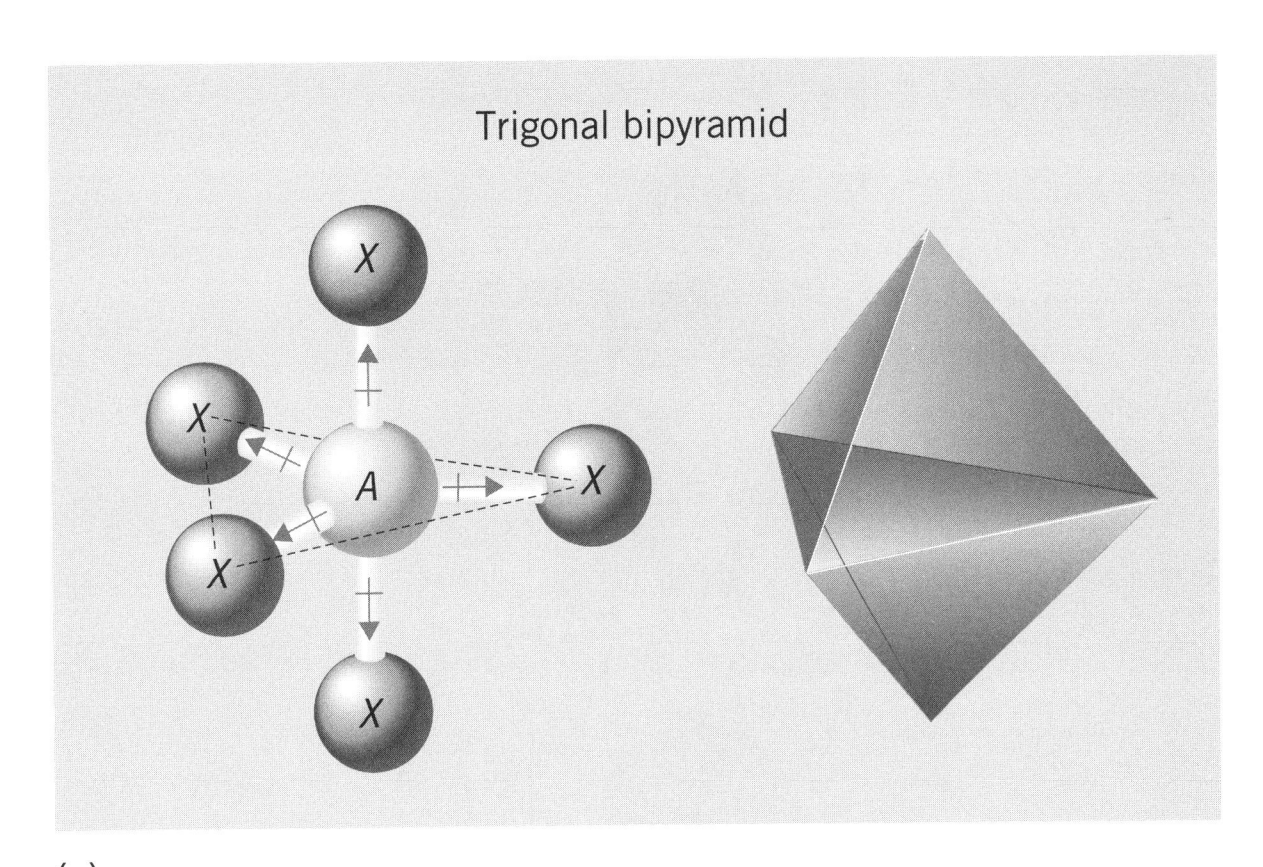

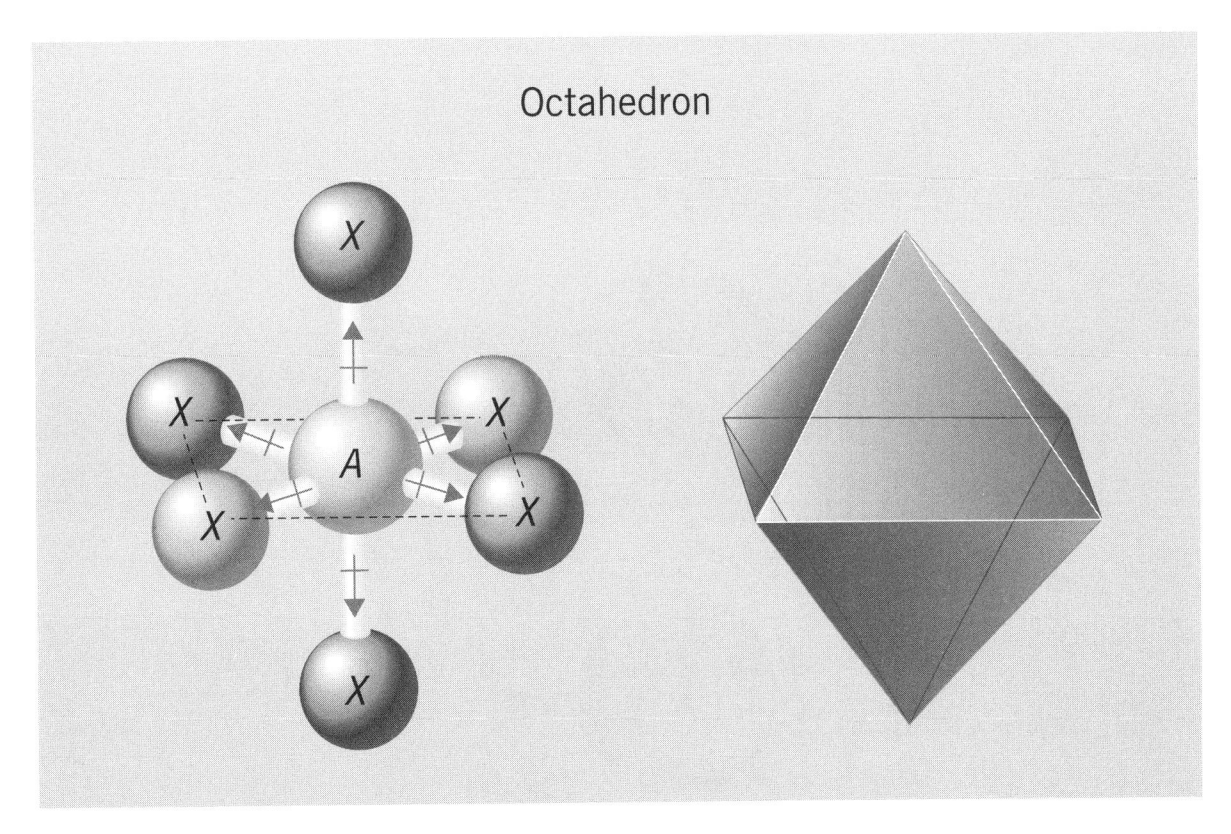

Trigonal bipyramid

(a)

Octahedron

(b)

FIGURE 9.8

Covalent bond in H₂

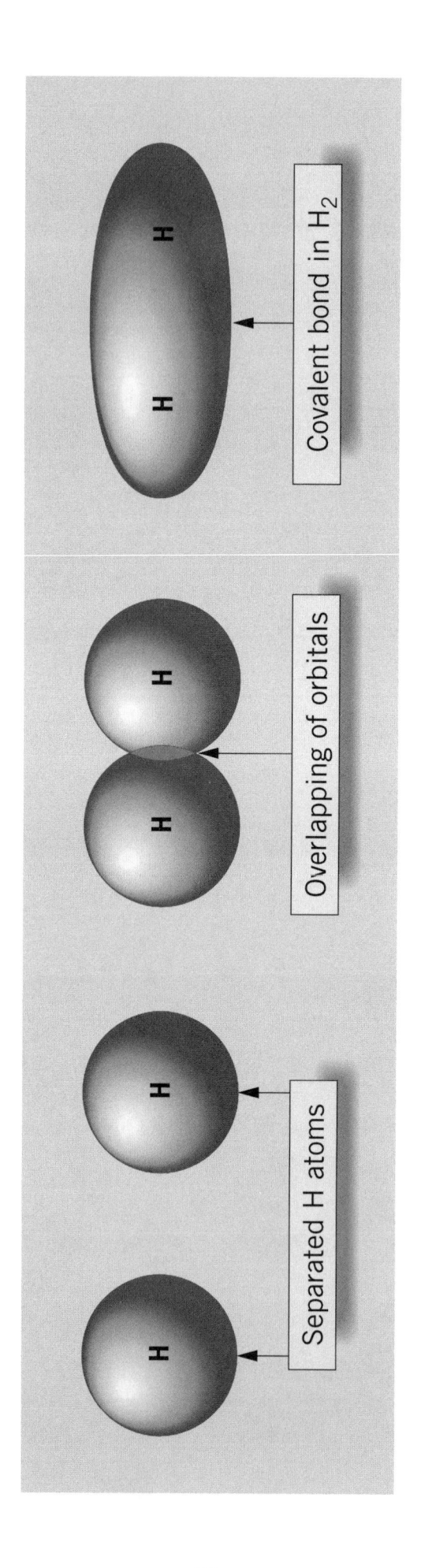

Overlapping of orbitals

Separated H atoms

FIGURE 9.11

FIGURE 9.12

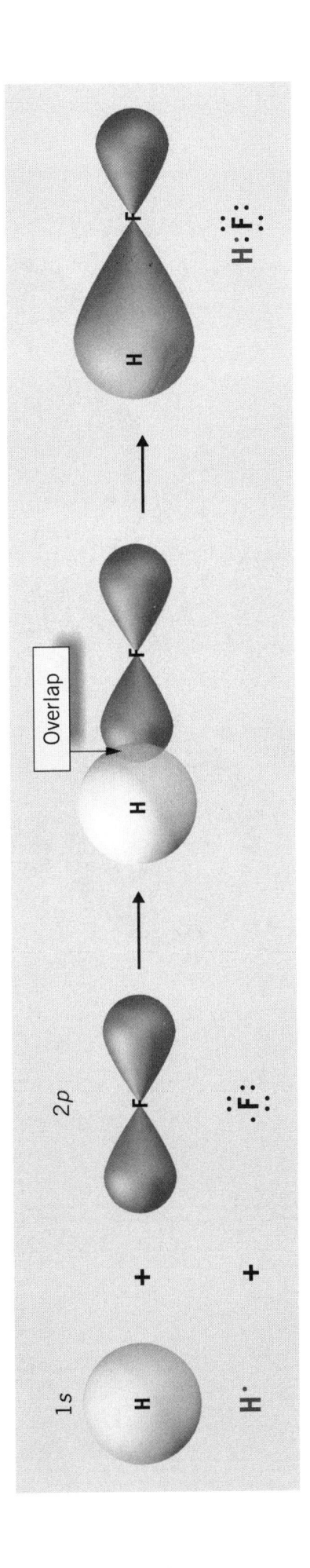

FIGURE 9.13

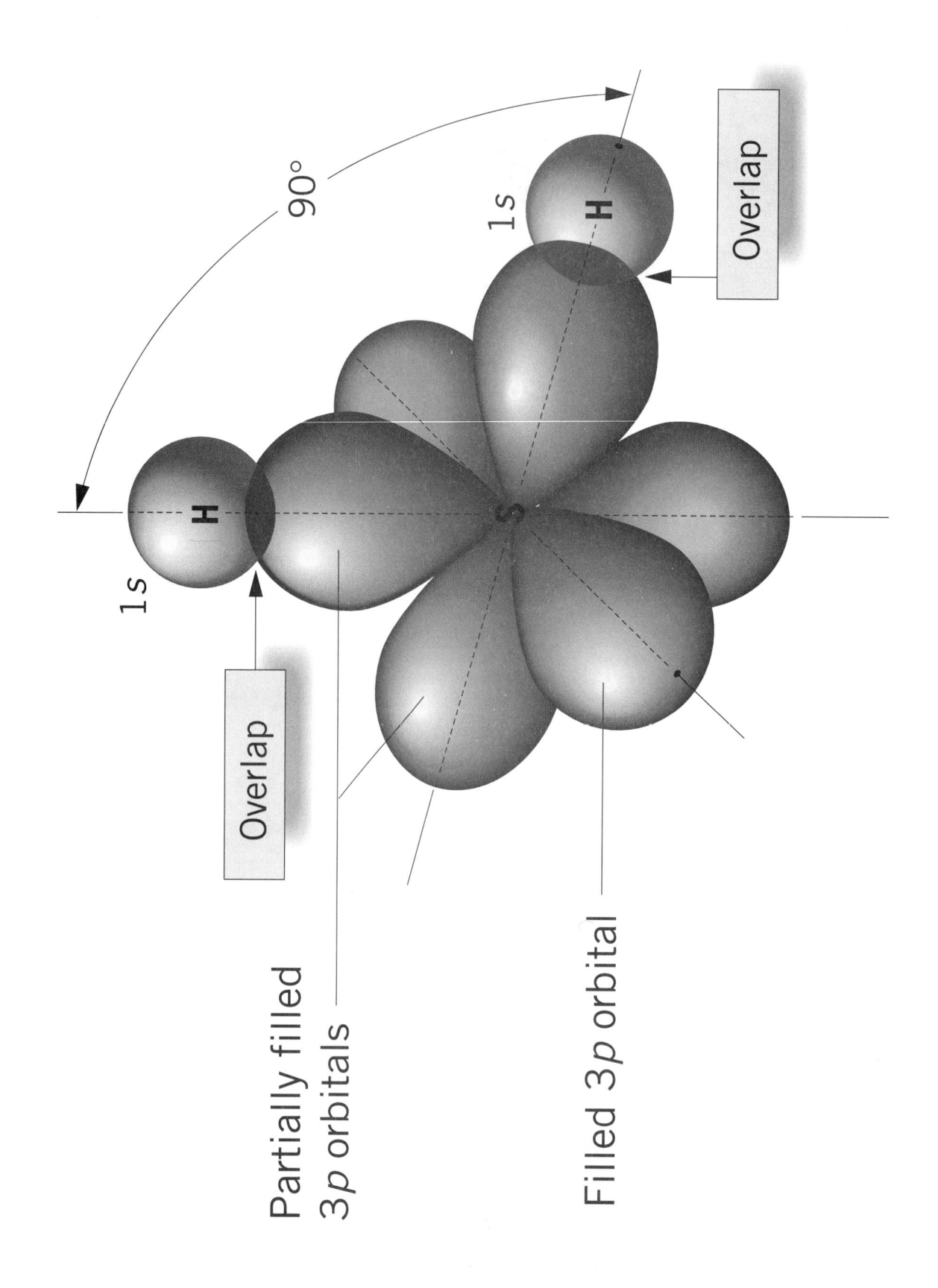

(a) 1s sp sp 1s

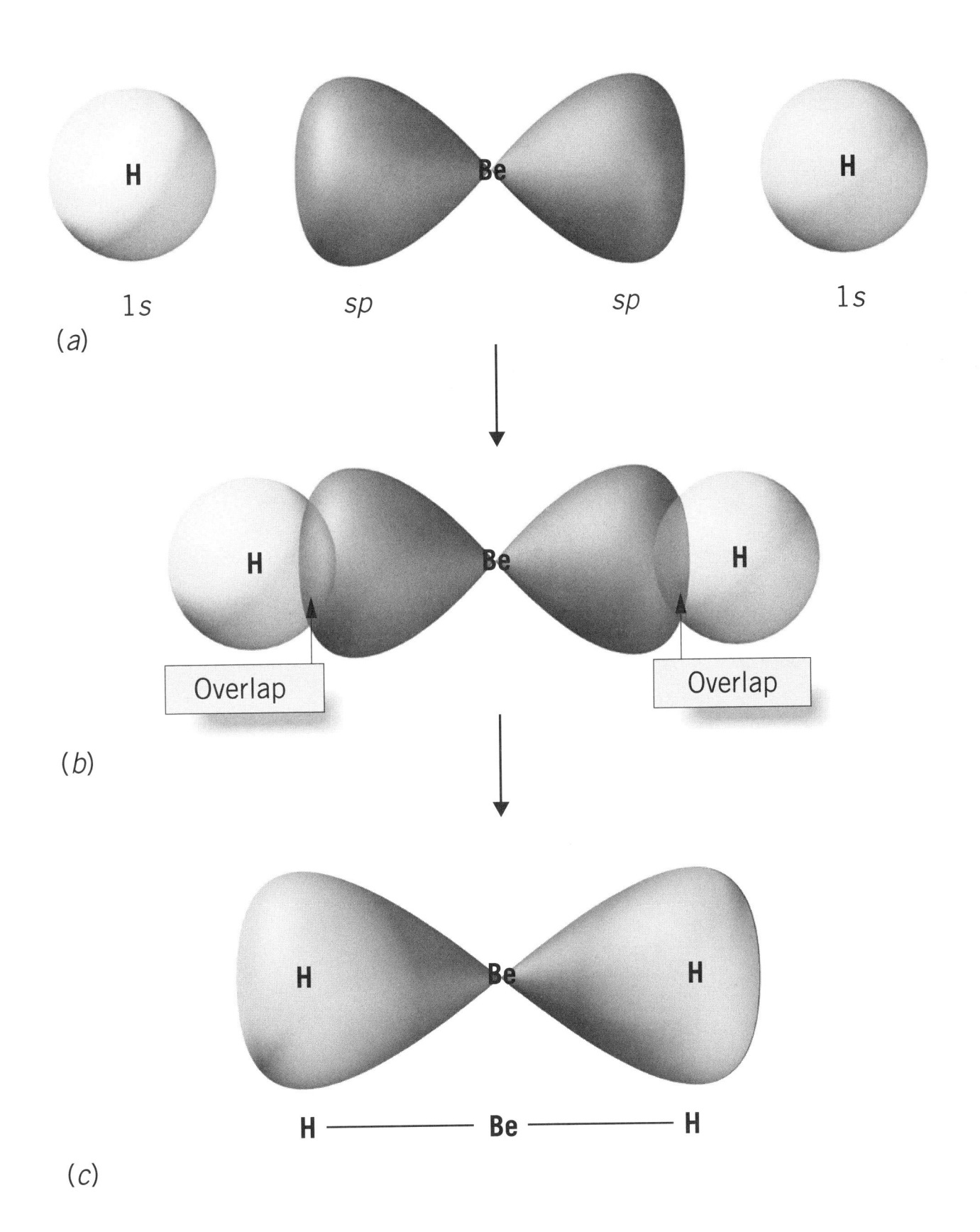

(b) Overlap Overlap

(c) H—Be—H

FIGURE 9.16

Linear

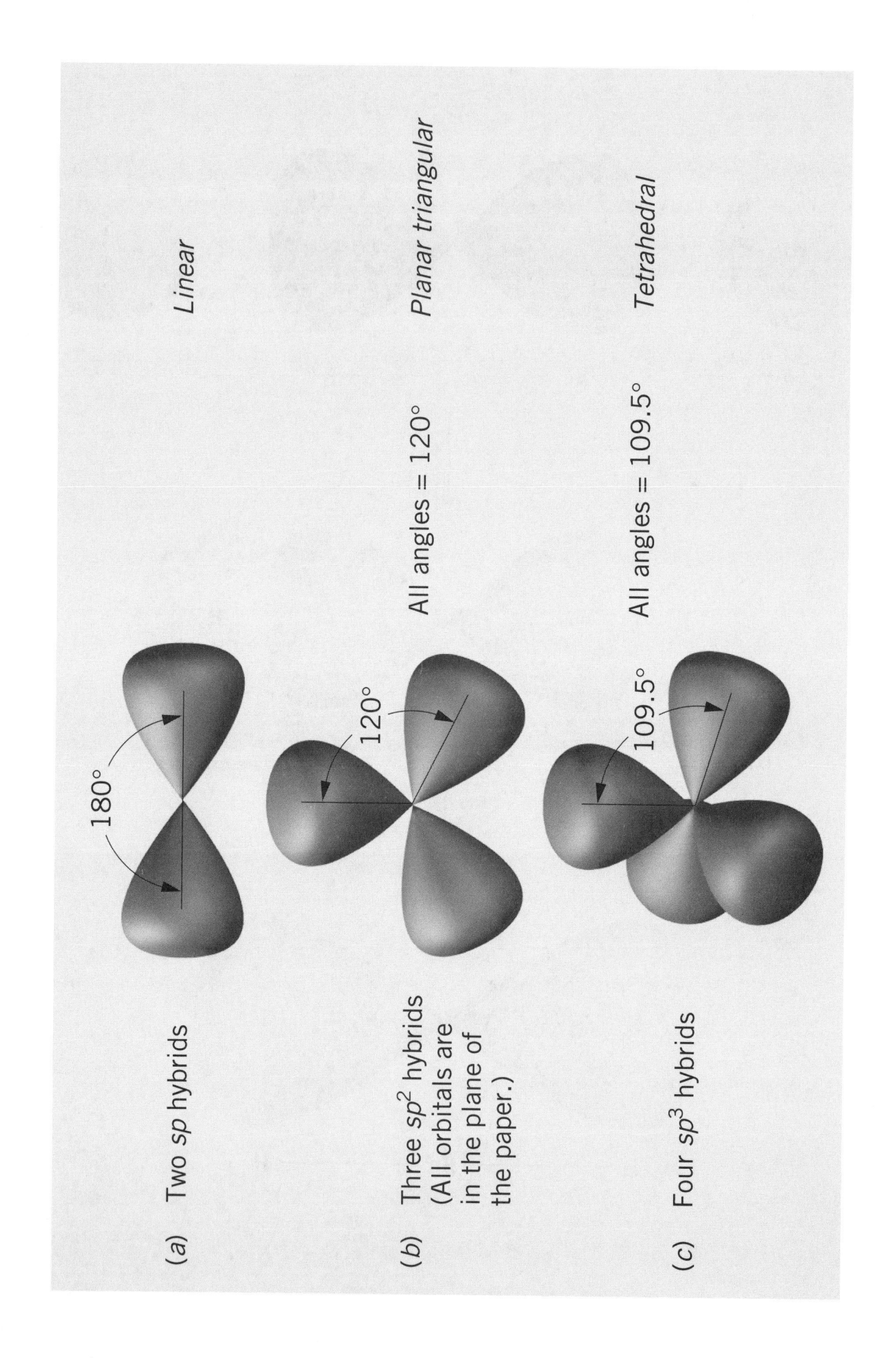

Planar triangular

All angles = 120°

Tetrahedral

All angles = 109.5°

180°

120°

109.5°

(a) Two *sp* hybrids

(b) Three *sp*² hybrids
(All orbitals are in the plane of the paper.)

(c) Four *sp*³ hybrids

FIGURE 9.17

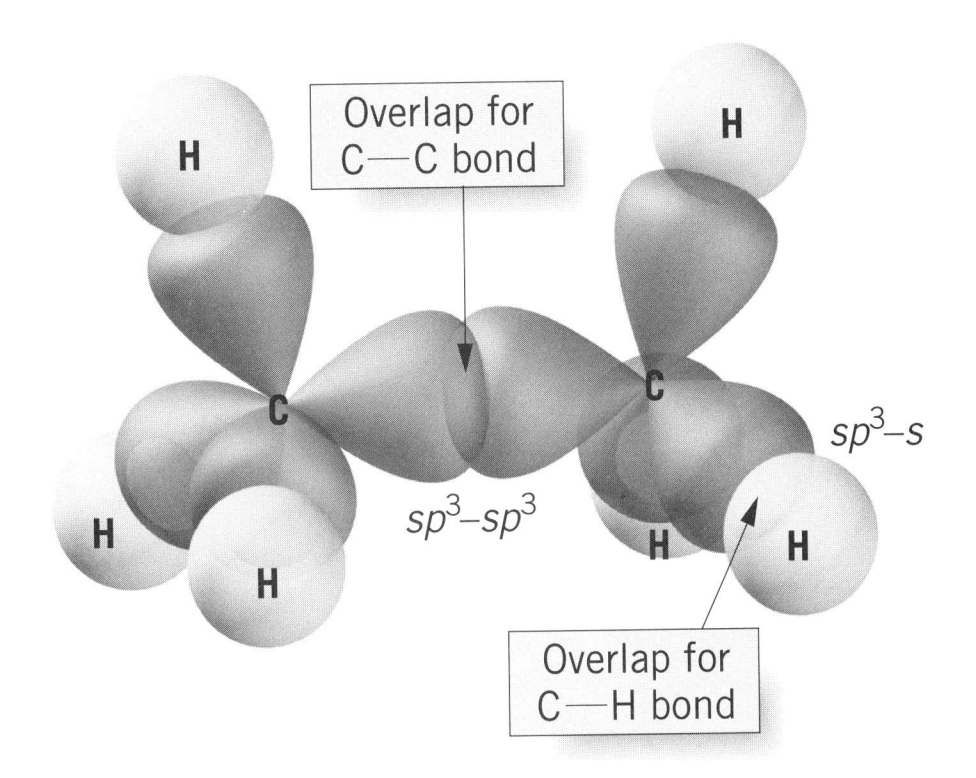

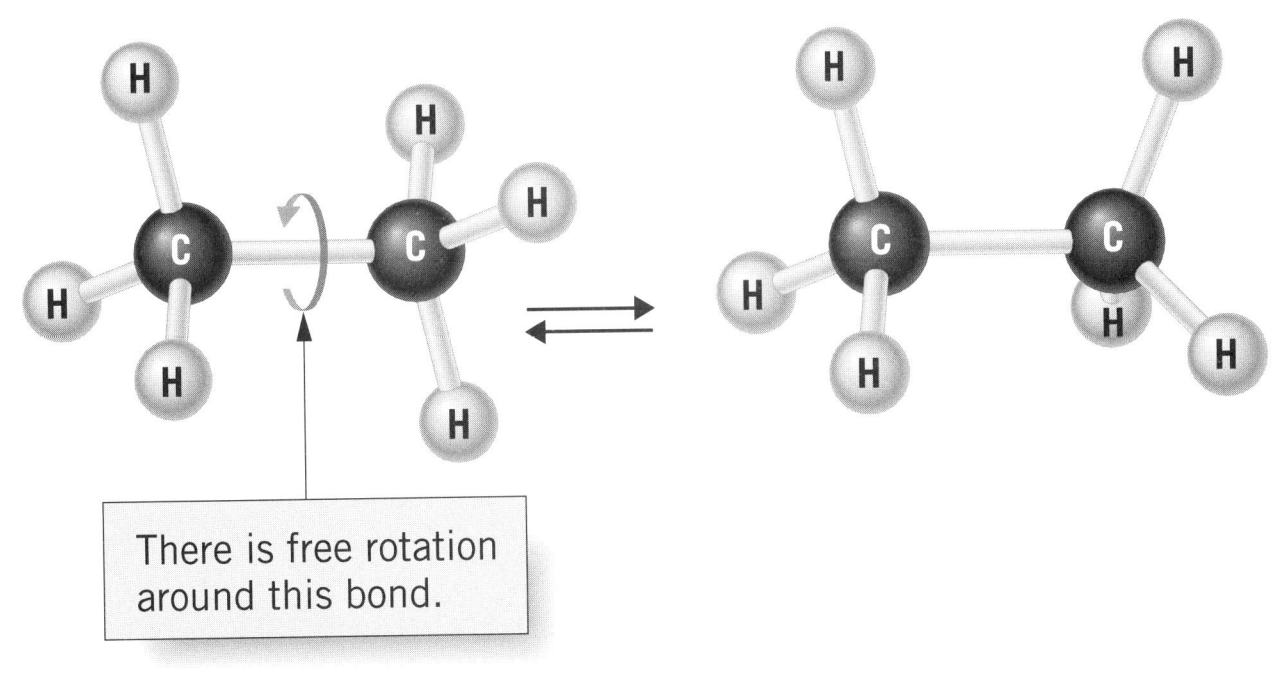

(a)

(b)

FIGURE 9.20

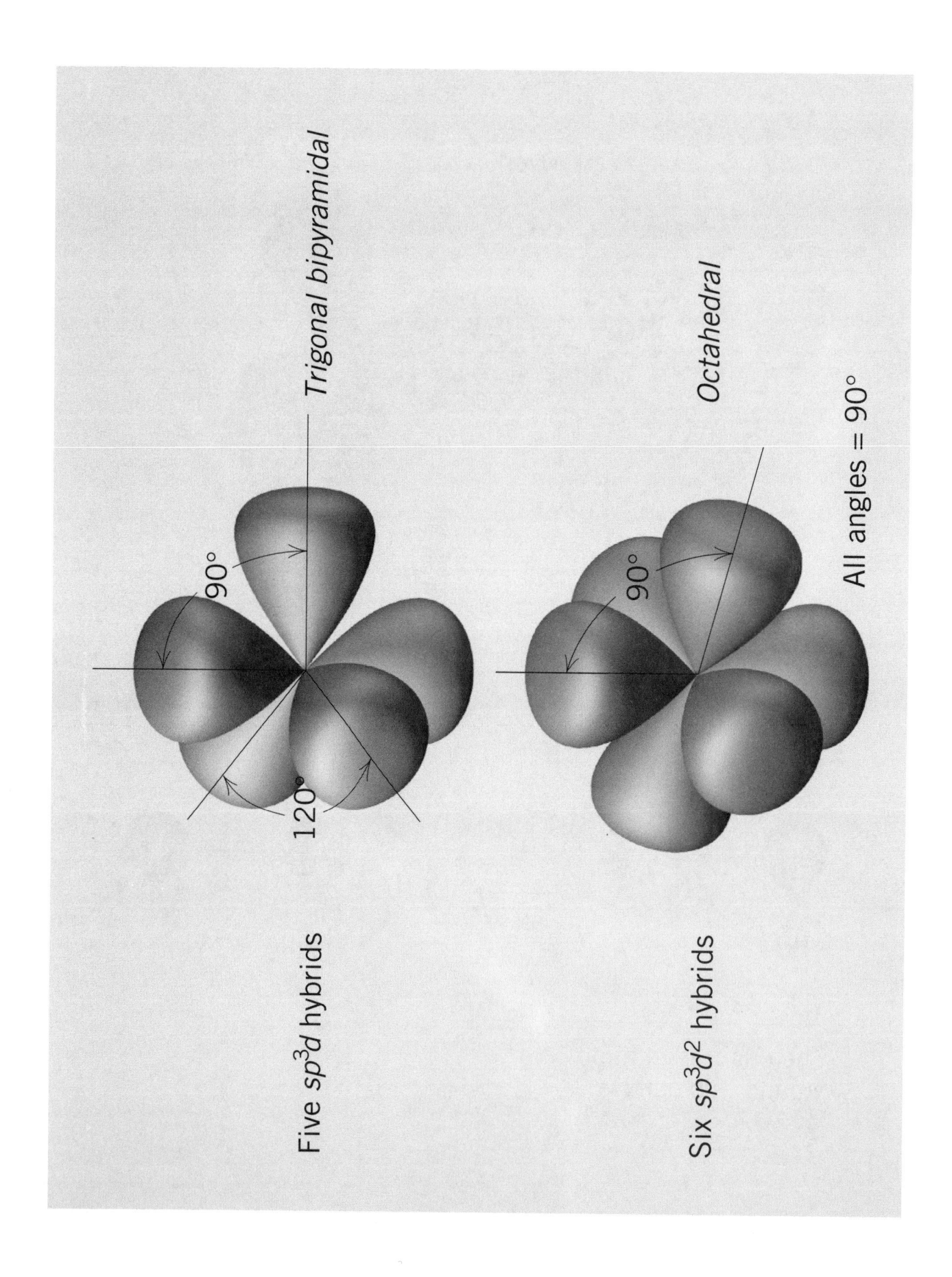

Trigonal bipyramidal

Octahedral

90°

120°

90°

All angles = 90°

Five *sp³d* hybrids

Six *sp³d²* hybrids

FIGURE 9.22

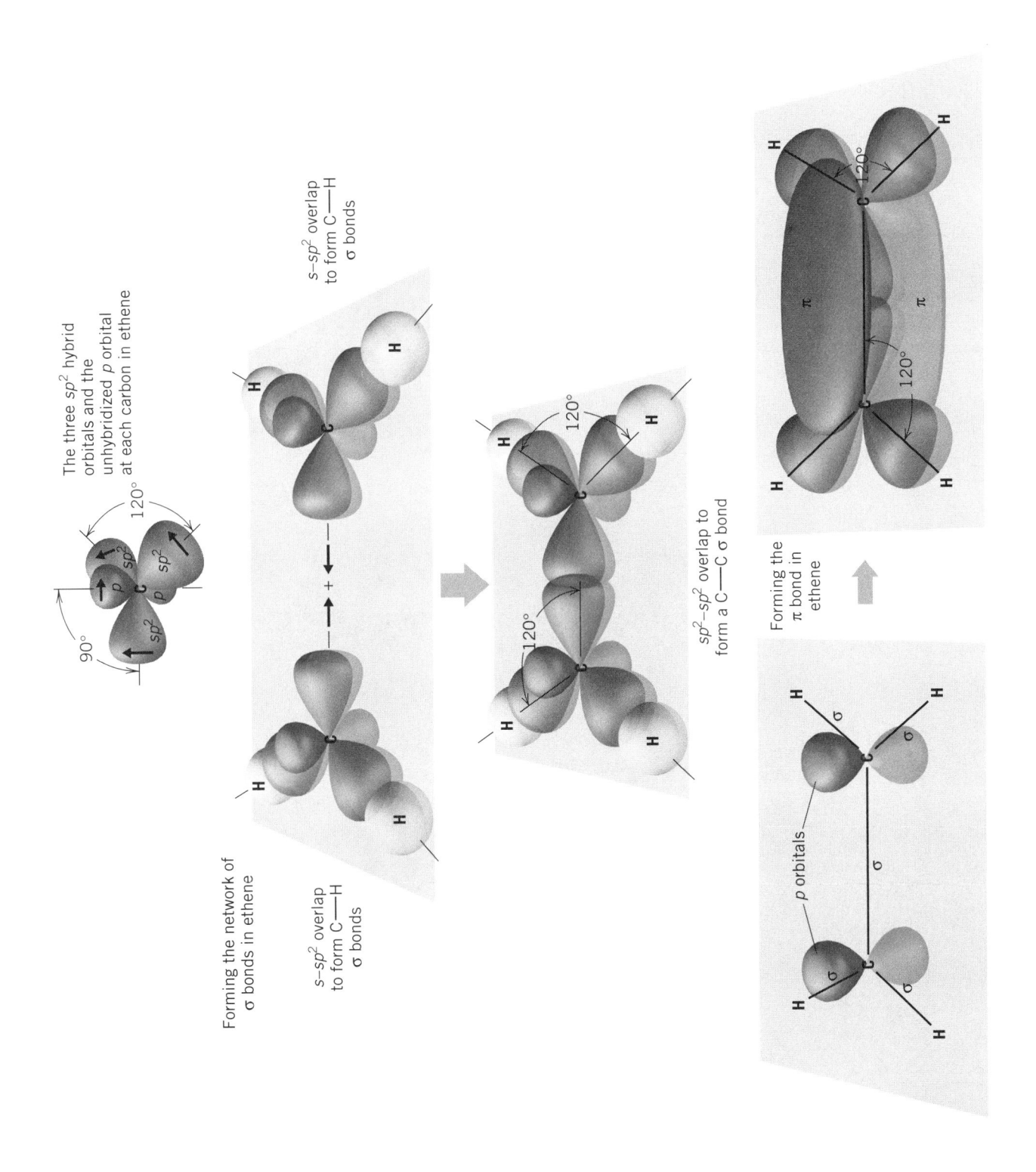

The three sp^2 hybrid orbitals and the unhybridized p orbital at each carbon in ethene

s–sp^2 overlap to form C—H σ bonds

s–sp^2 overlap to form C—H σ bonds

Forming the network of σ bonds in ethene

sp^2–sp^2 overlap to form a C—C σ bond

Forming the π bond in ethene

FIGURE 9.25

H—C≡C—H
acetylene

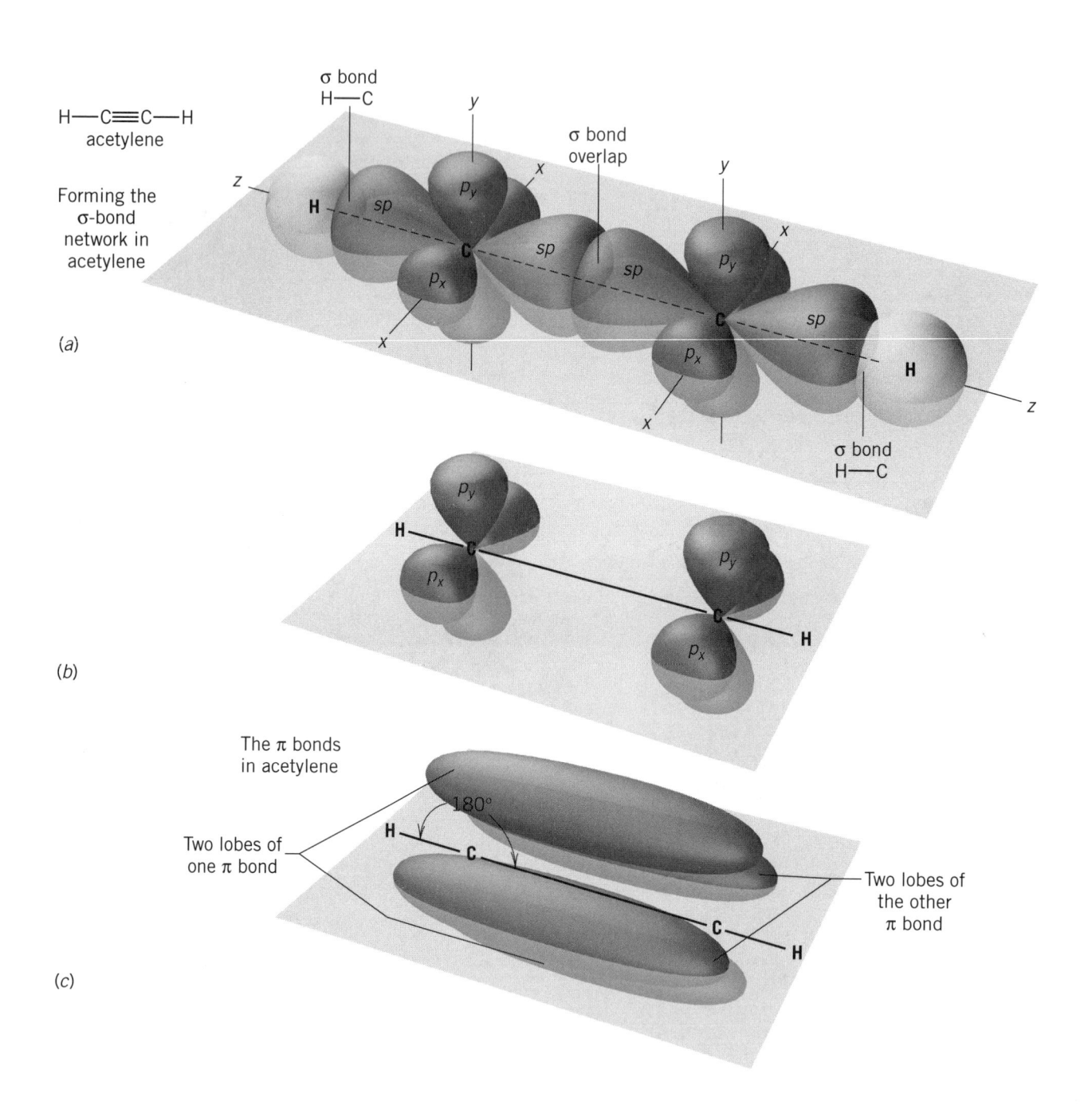

Forming the σ-bond network in acetylene

(a) σ bond H—C; σ bond overlap; σ bond H—C

(b)

(c) The π bonds in acetylene
Two lobes of one π bond
180°
Two lobes of the other π bond

FIGURE 9.28

FIGURE 9.33

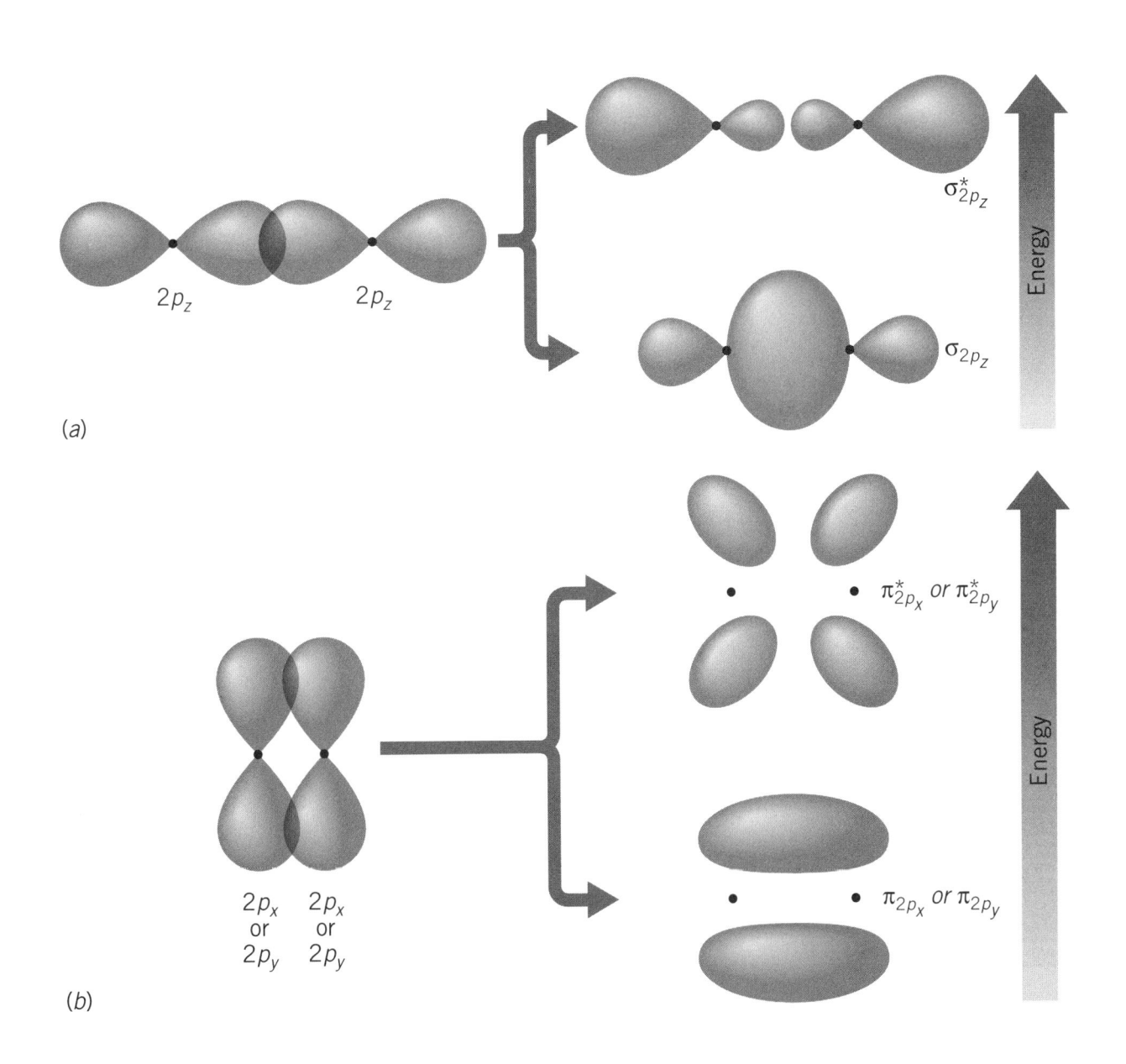

FIGURE 9.34

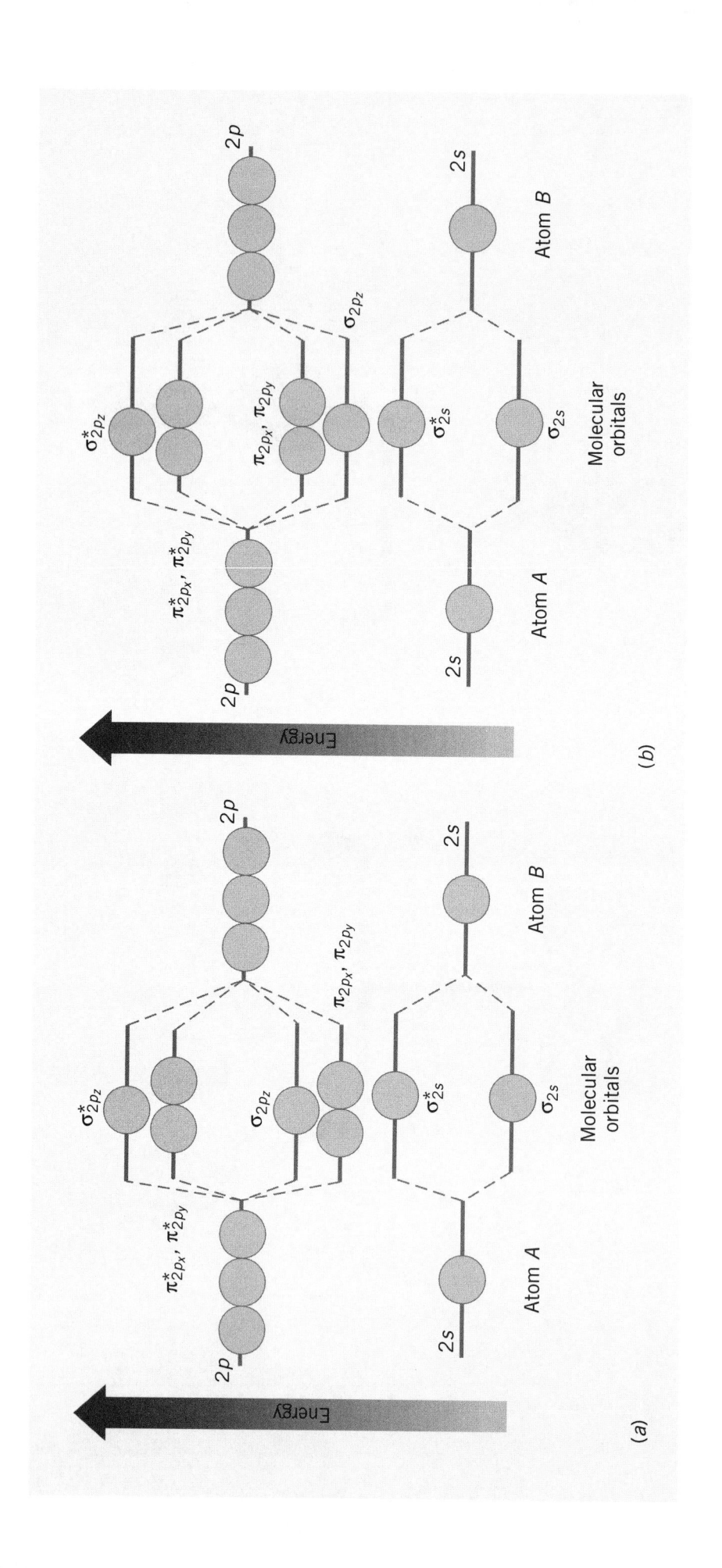

FIGURE 9.38

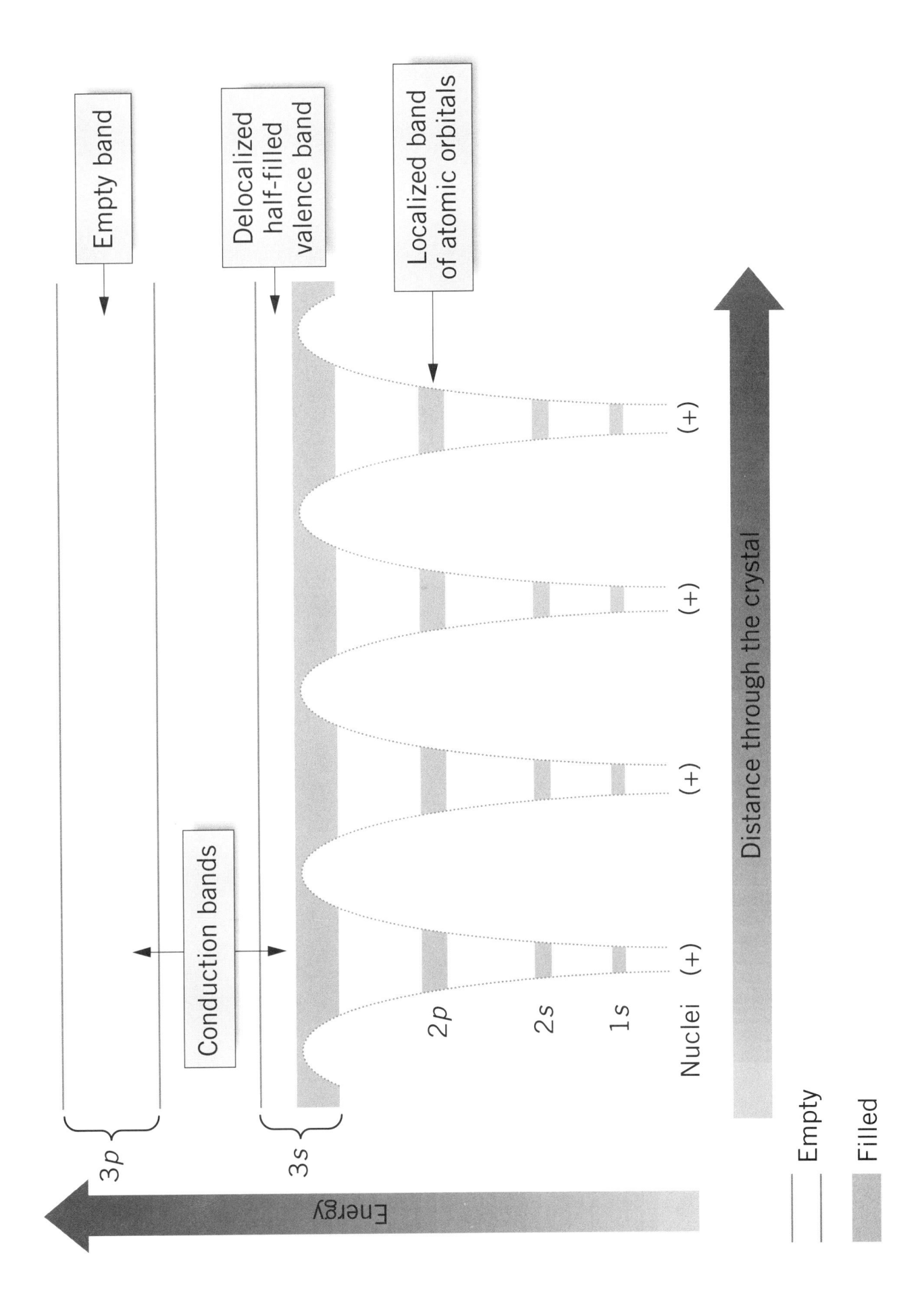

Vacuum (no weight of air here)

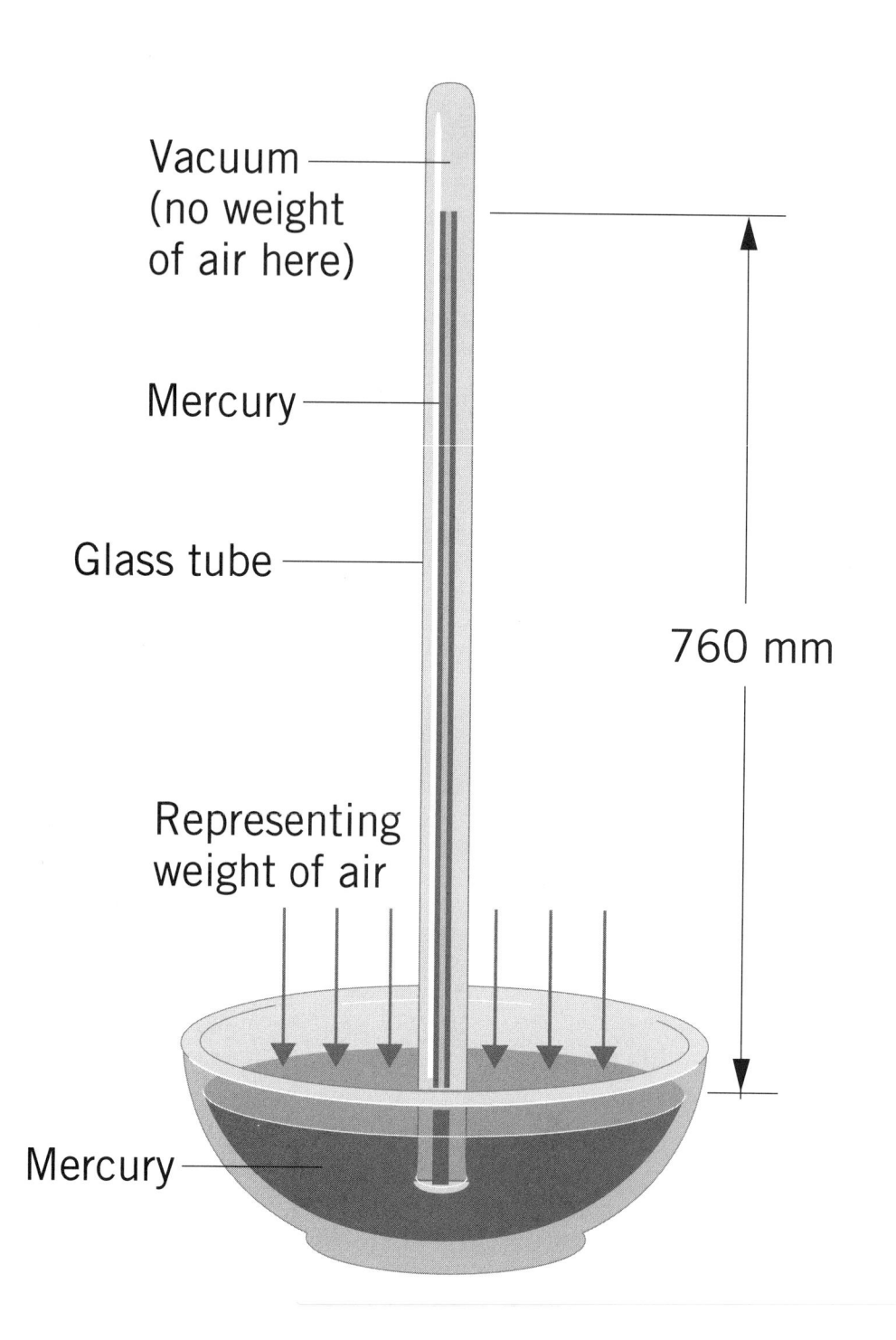

Mercury

Glass tube

760 mm

Representing weight of air

Mercury

FIGURE 10.2

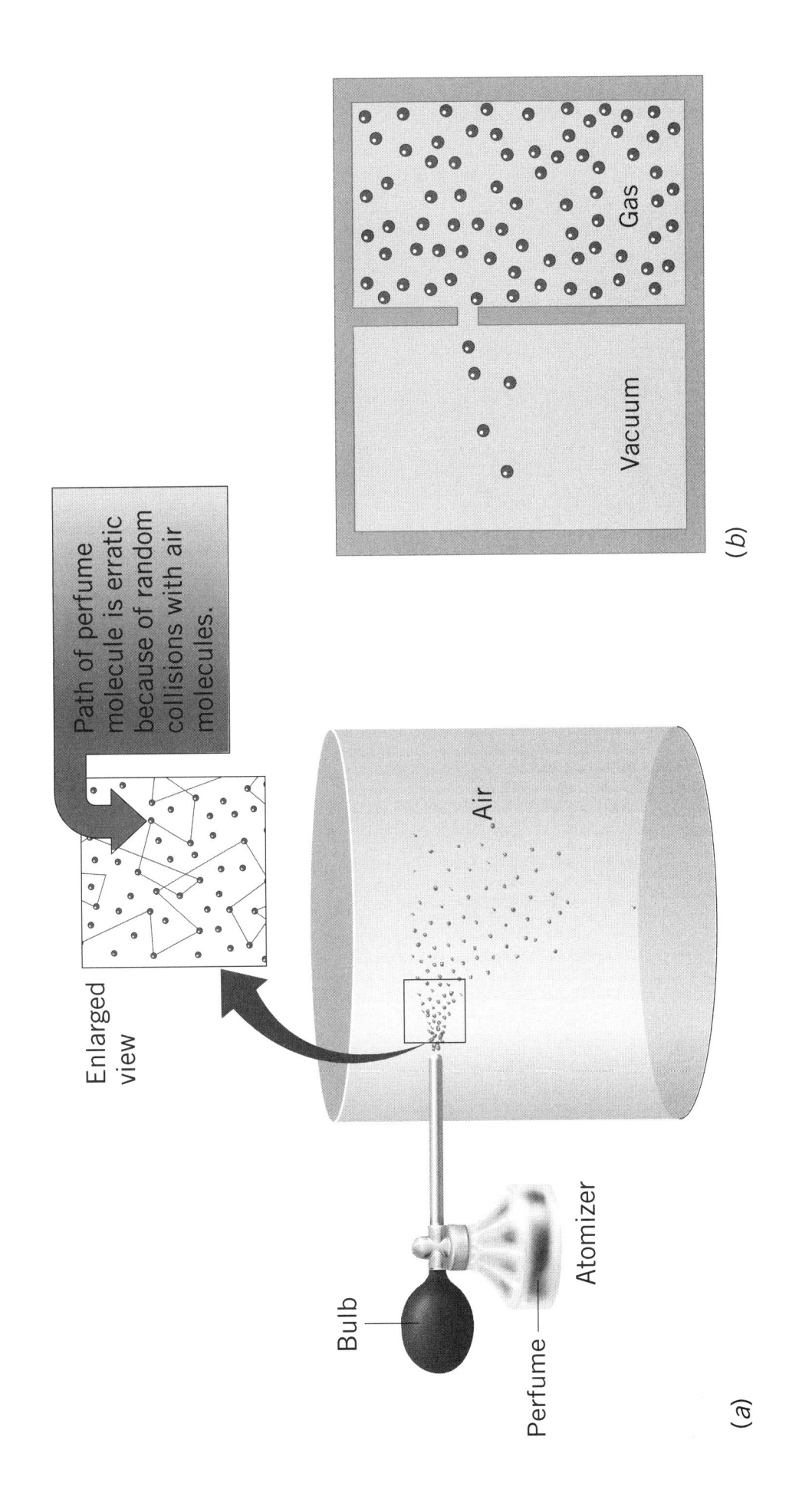

Path of perfume molecule is erratic because of random collisions with air molecules.

Enlarged view

Air

Bulb

Perfume

Atomizer

(a)

Gas

Vacuum

(b)

FIGURE 10.10

(a) Lower pressure
1 kg

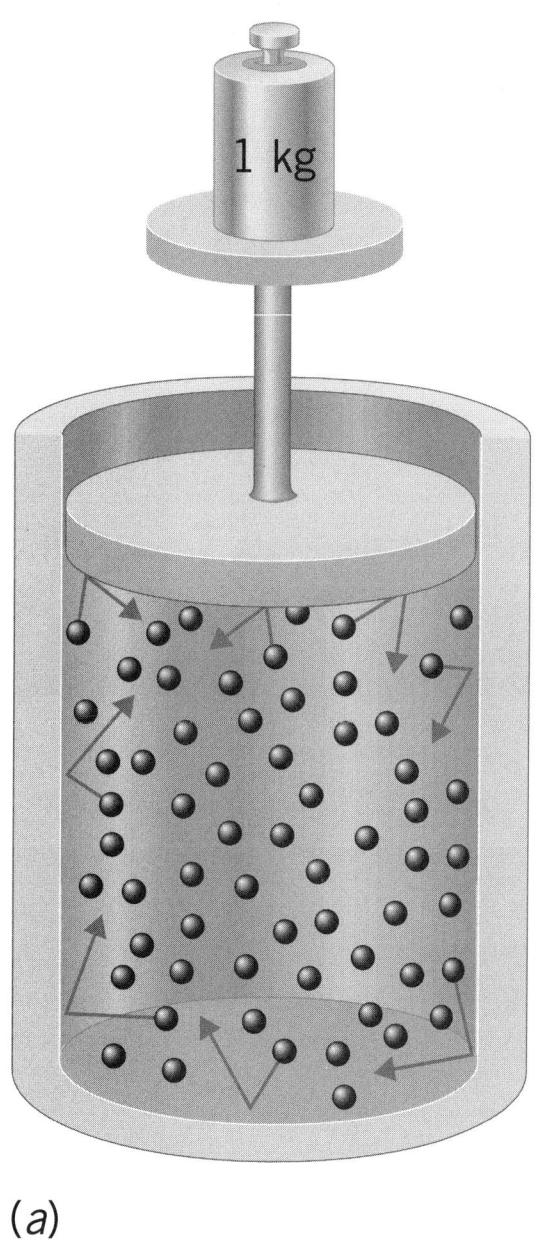

(b) Higher pressure
1 kg 1 kg

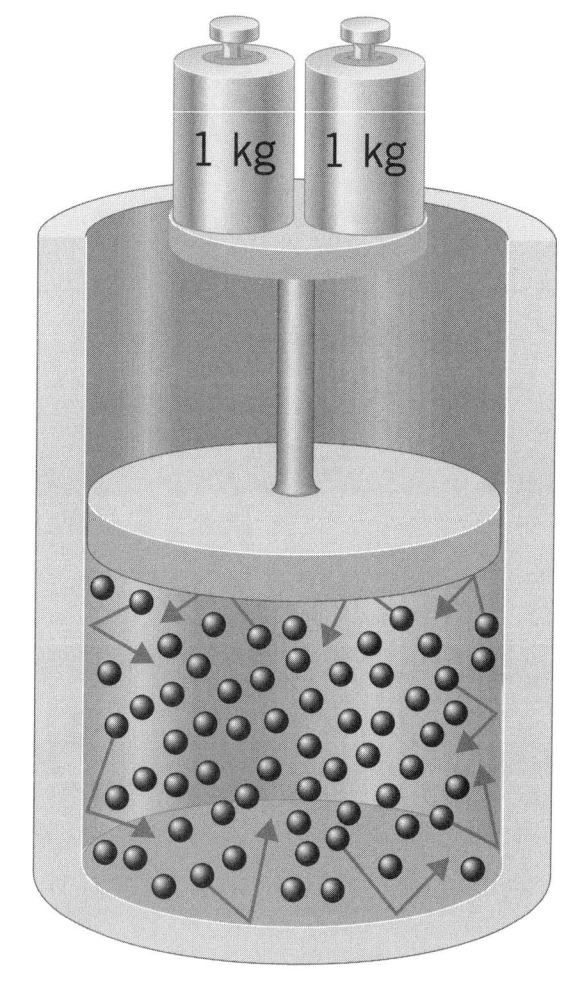

FIGURE 10.12

FIGURE 10.13

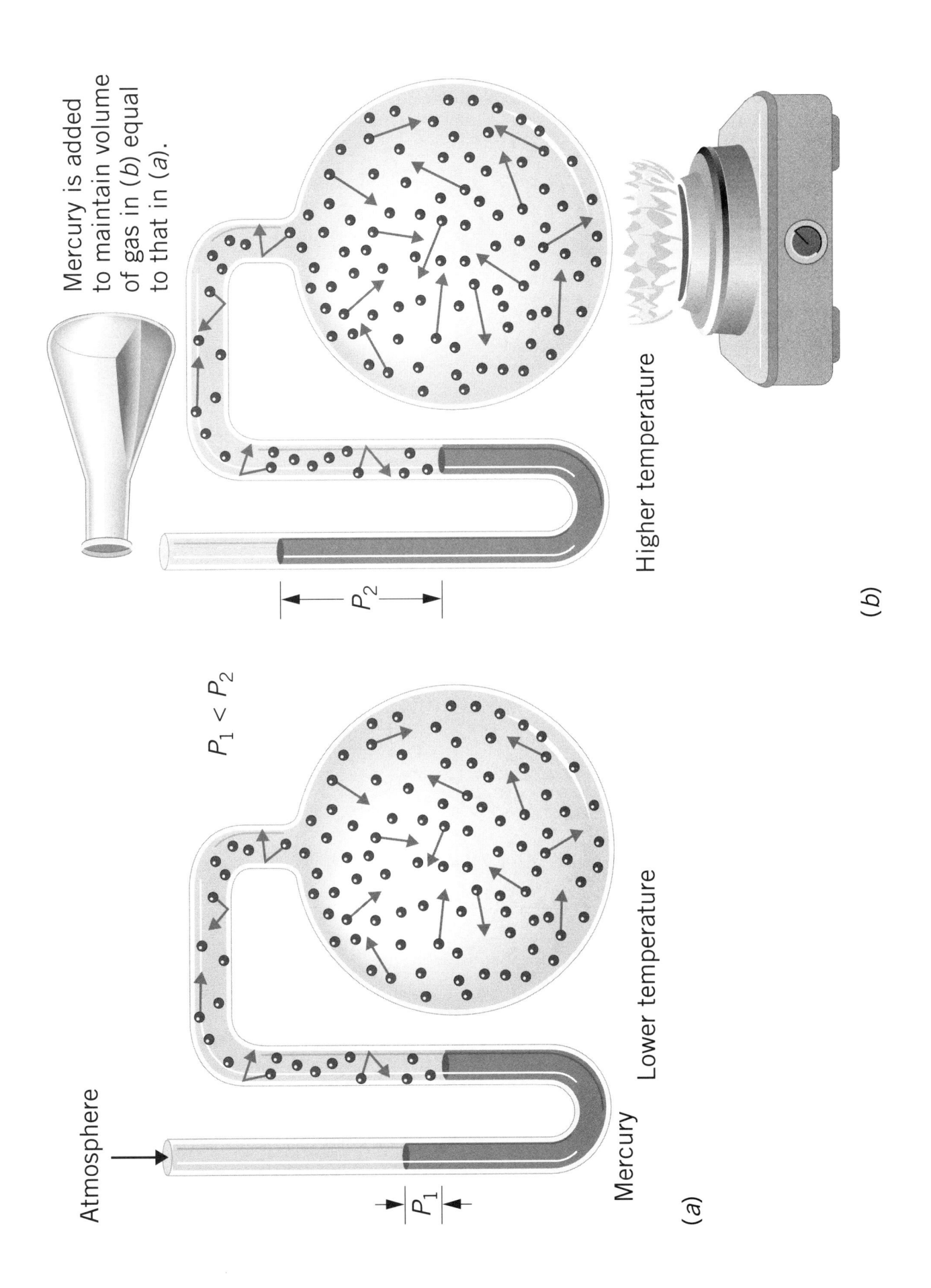

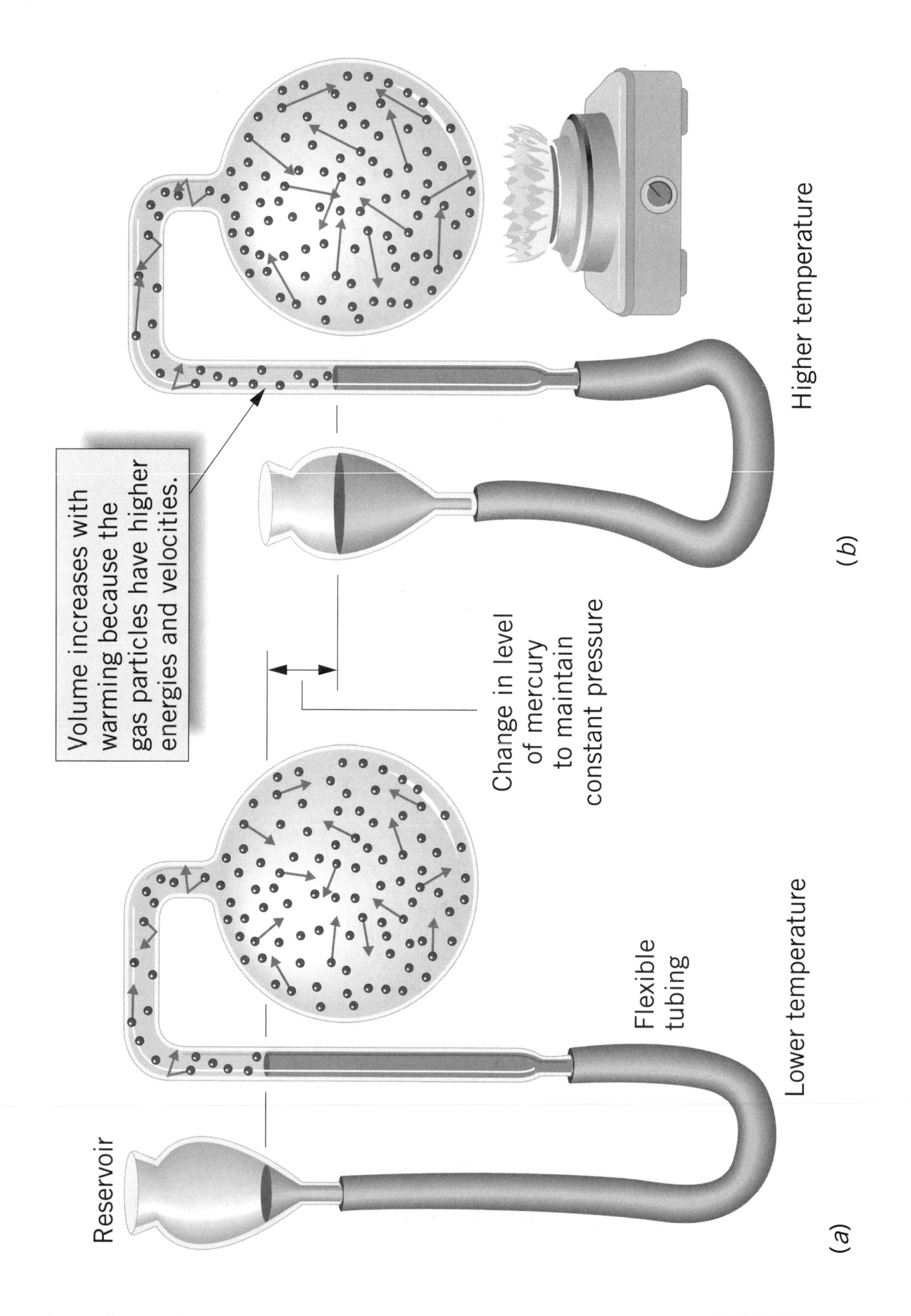

Volume increases with warming because the gas particles have higher energies and velocities.

Change in level of mercury to maintain constant pressure

Reservoir

Flexible tubing

Lower temperature

(a)

Higher temperature

(b)

FIGURE 10.14

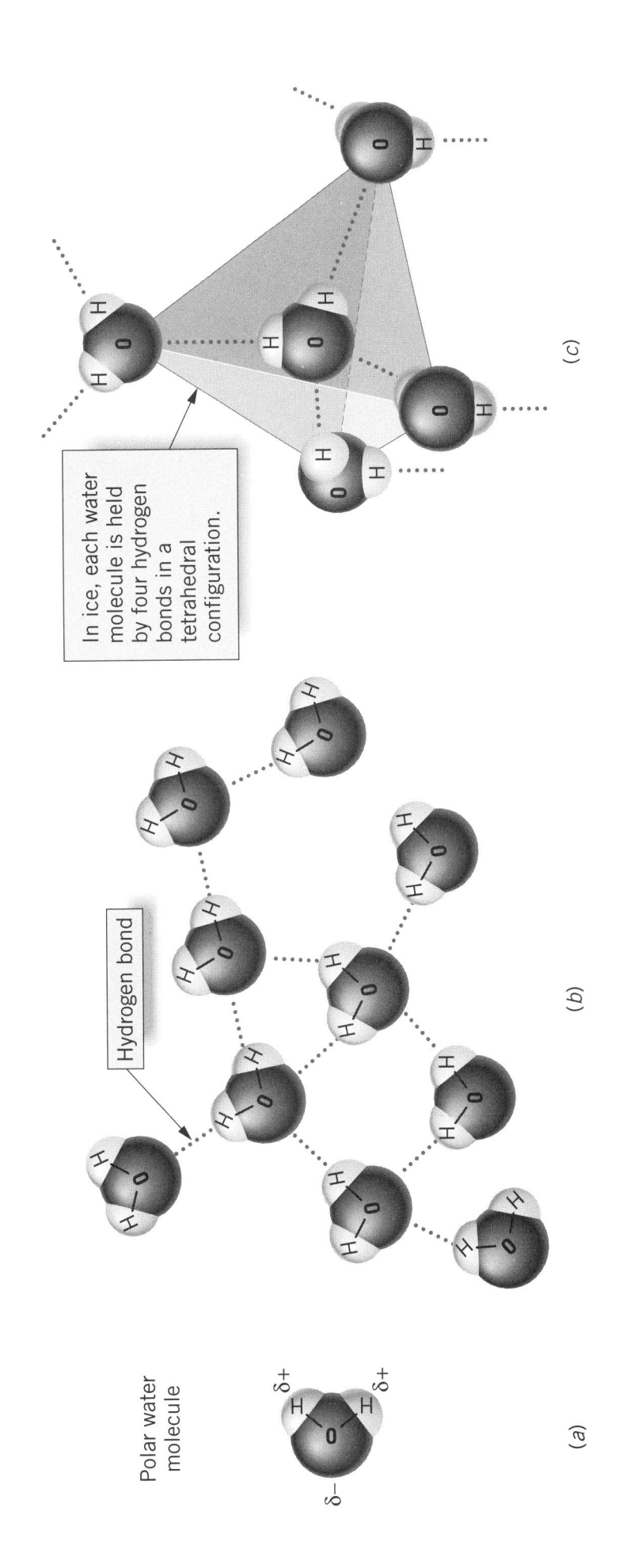

In ice, each water molecule is held by four hydrogen bonds in a tetrahedral configuration.

(c)

Hydrogen bond

(b)

Polar water molecule

$\delta+$ $\delta+$
$\delta-$

(a)

FIGURE 11.3

Instantaneous dipoles

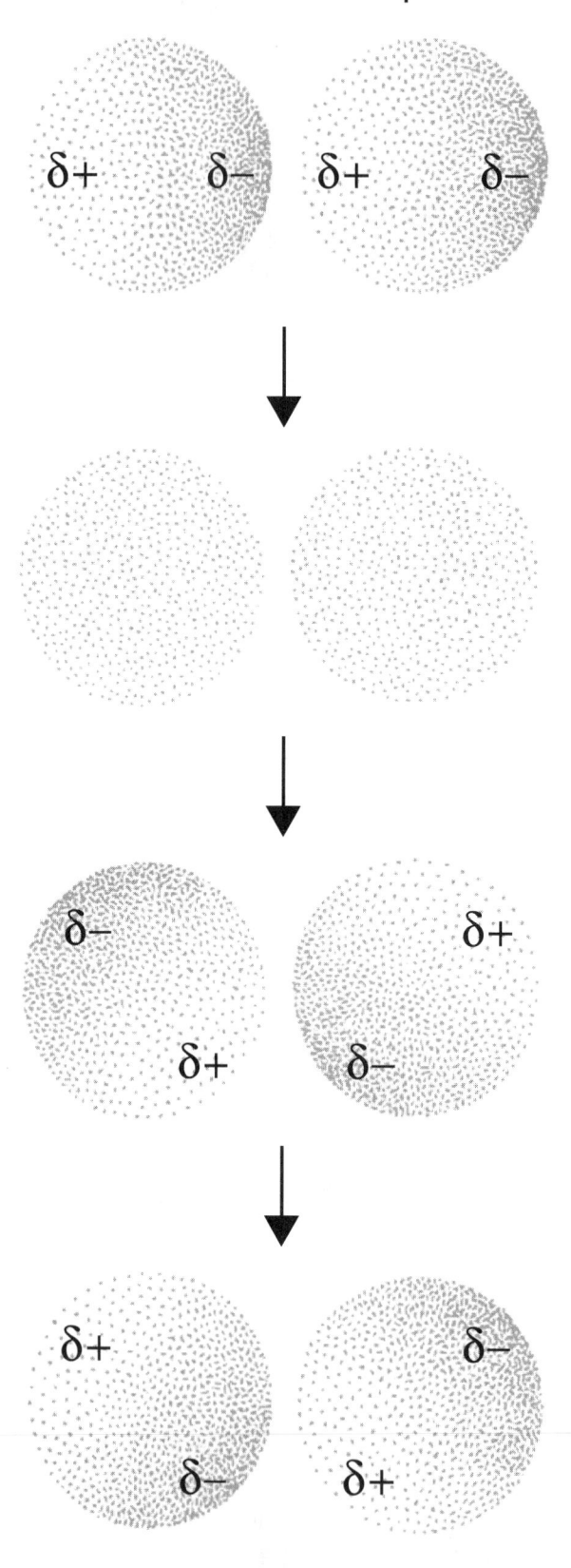

FIGURE 11.4

FIGURE 11.6

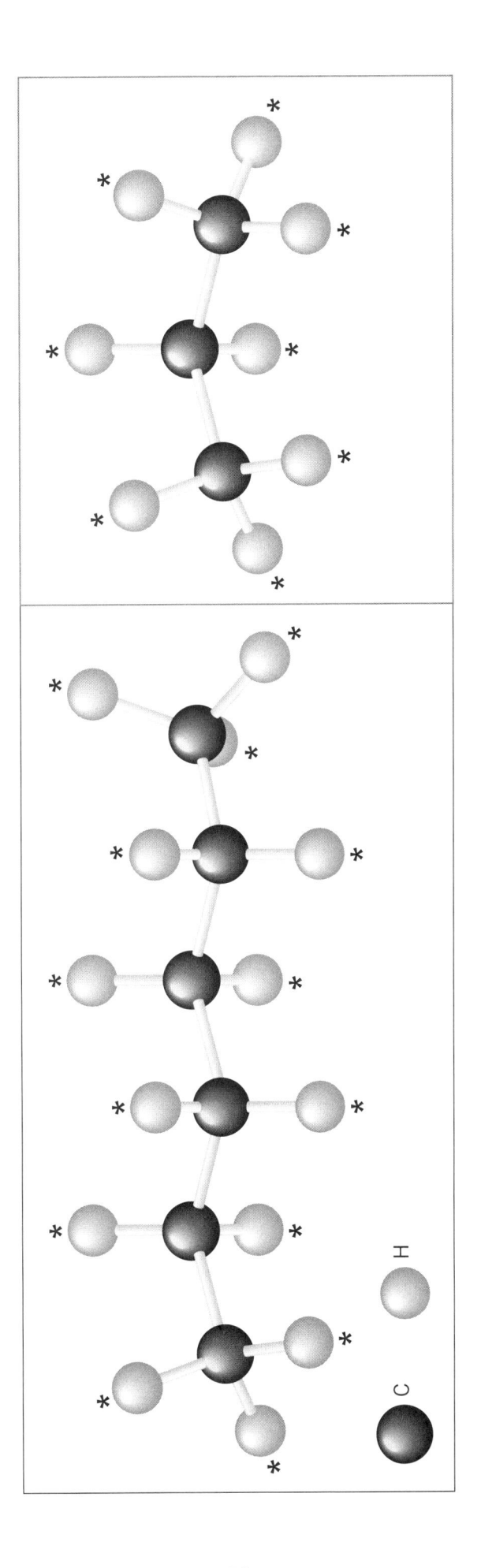

n-pentane, CH$_3$CH$_2$CH$_2$CH$_2$CH$_3$
bp = 36.1 °C

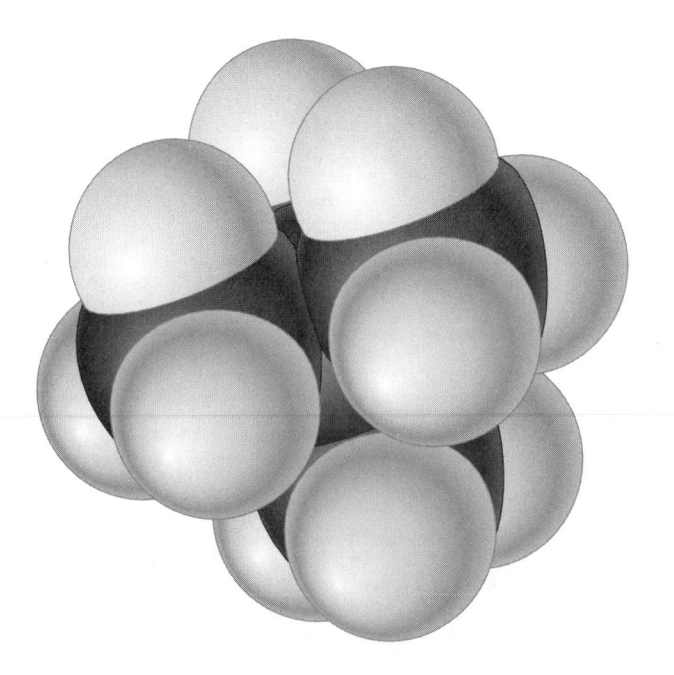

neopentane, (CH$_3$)$_4$C
bp = 9.5 °C

FIGURE 11.7

FIGURE 11.8

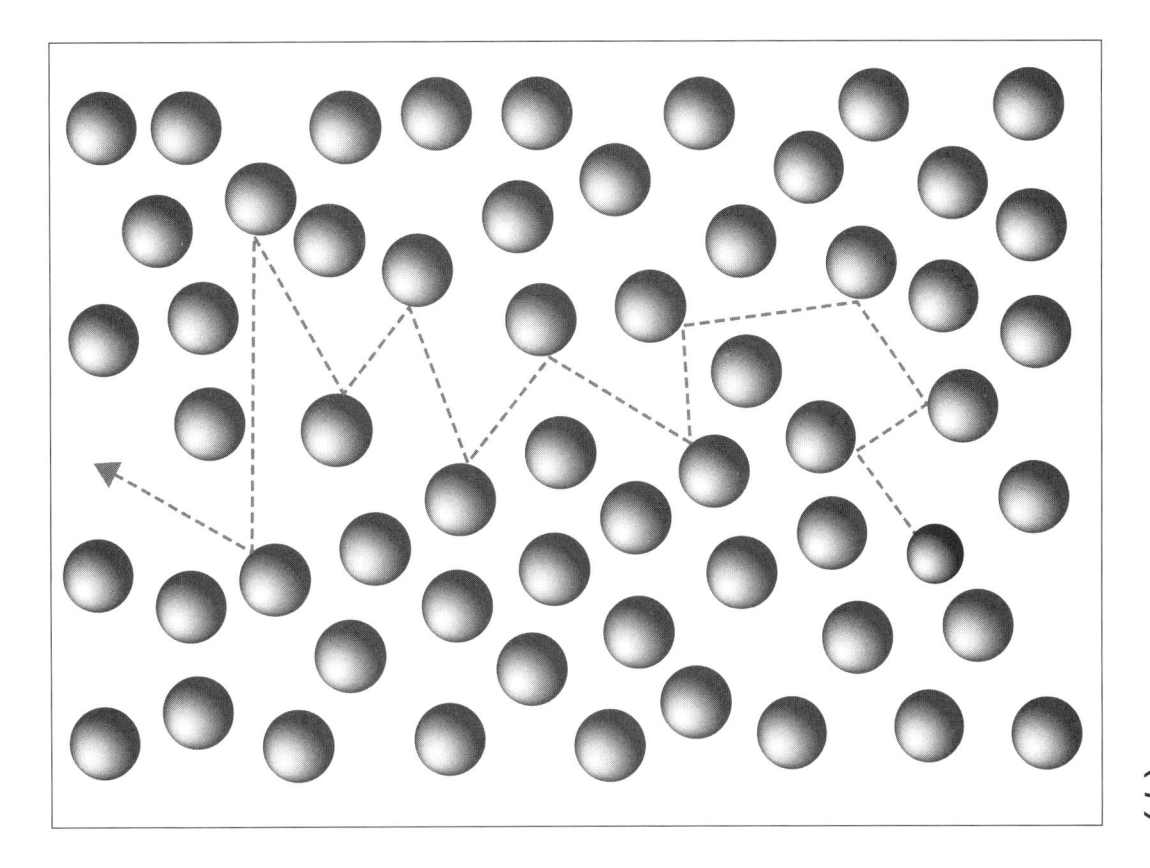

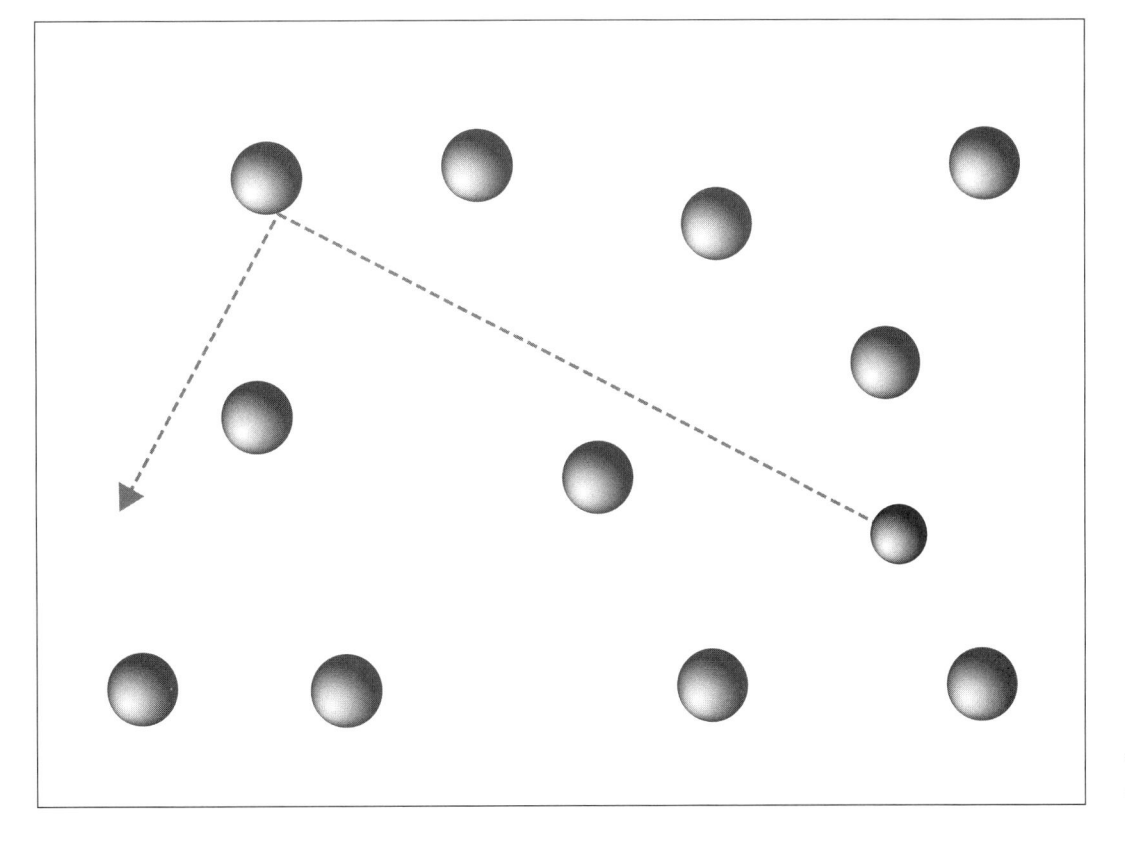

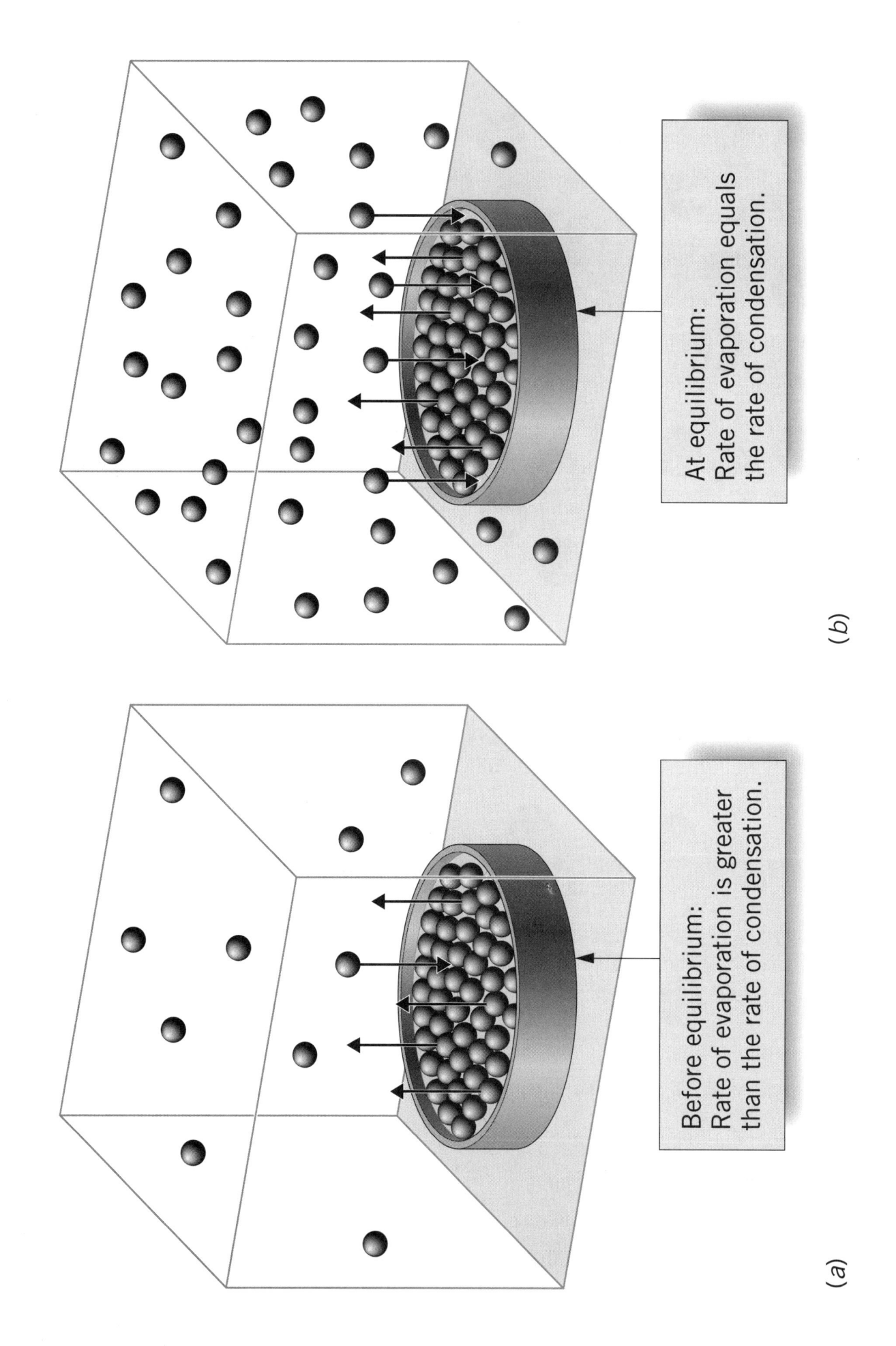

Before equilibrium: Rate of evaporation is greater than the rate of condensation.

(a)

At equilibrium: Rate of evaporation equals the rate of condensation.

(b)

FIGURE 11.15

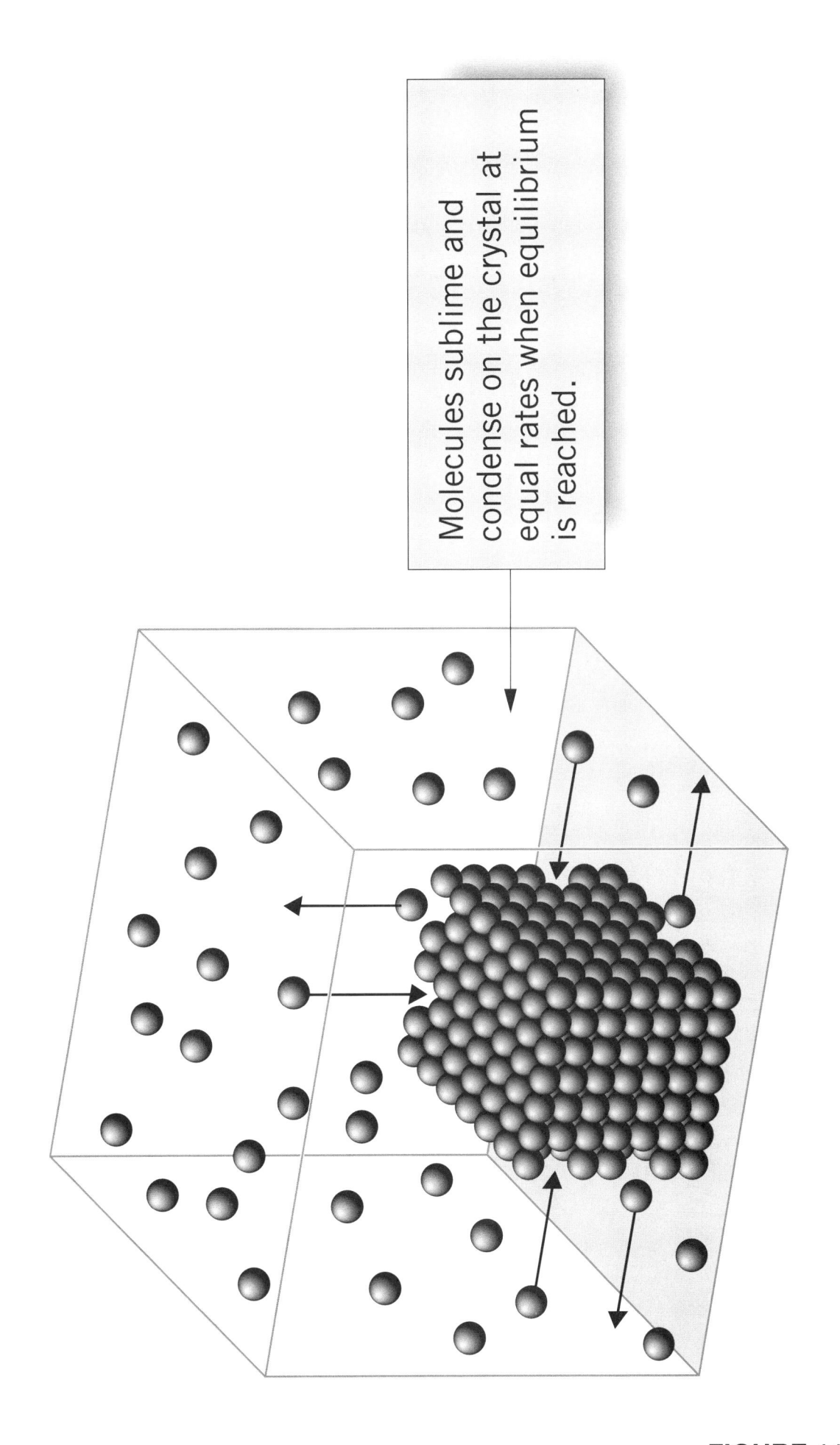

Molecules sublime and condense on the crystal at equal rates when equilibrium is reached.

FIGURE 11.16

FIGURE 11.17

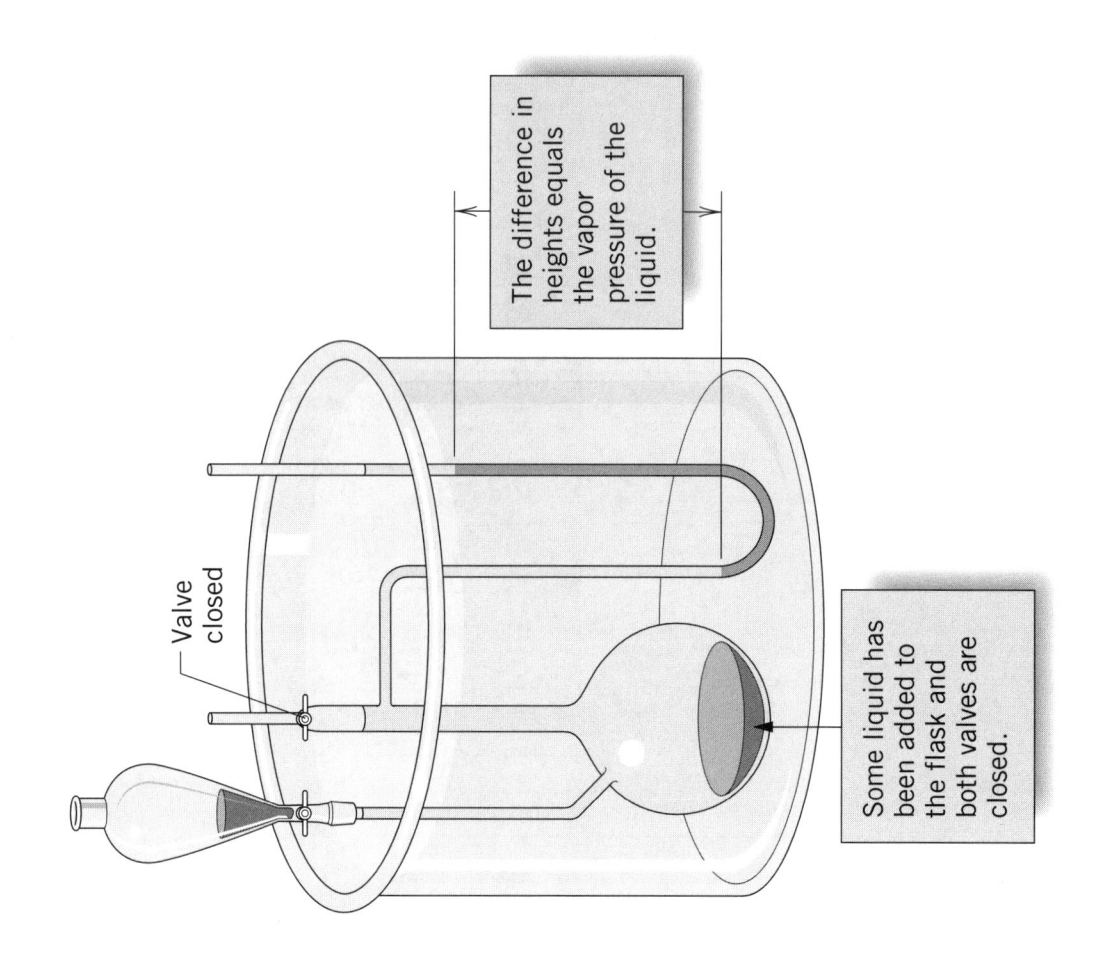

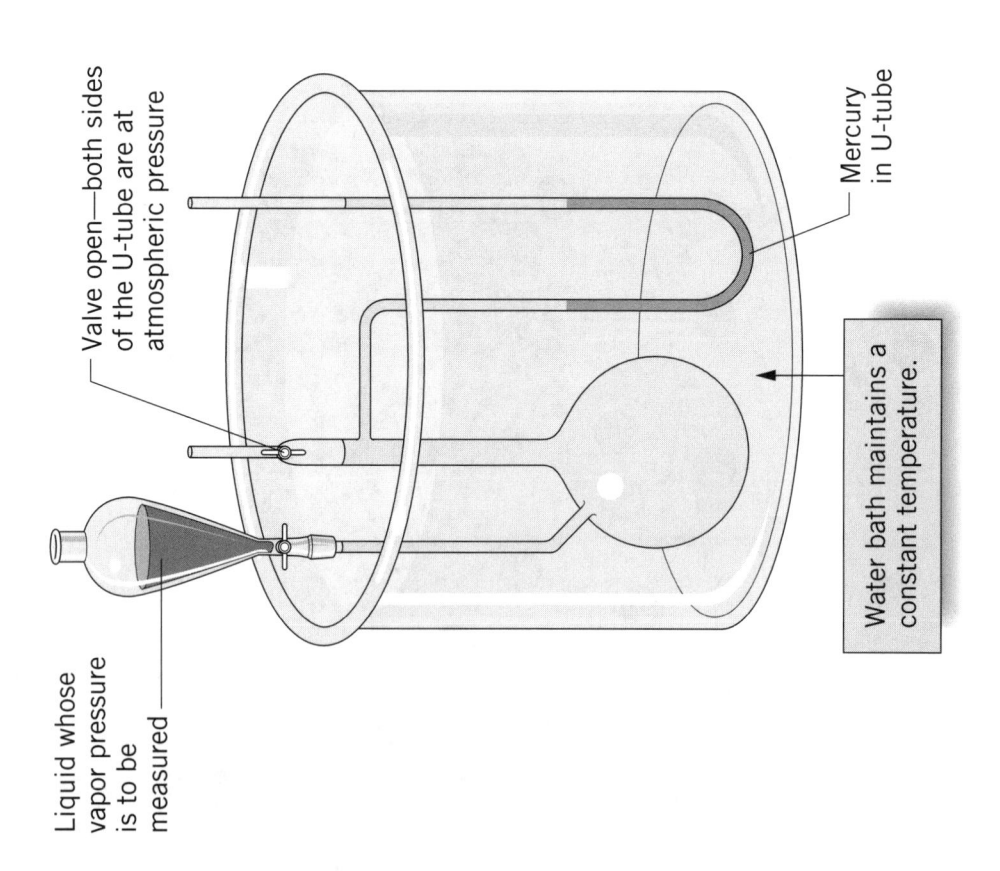

FIGURE 11.19

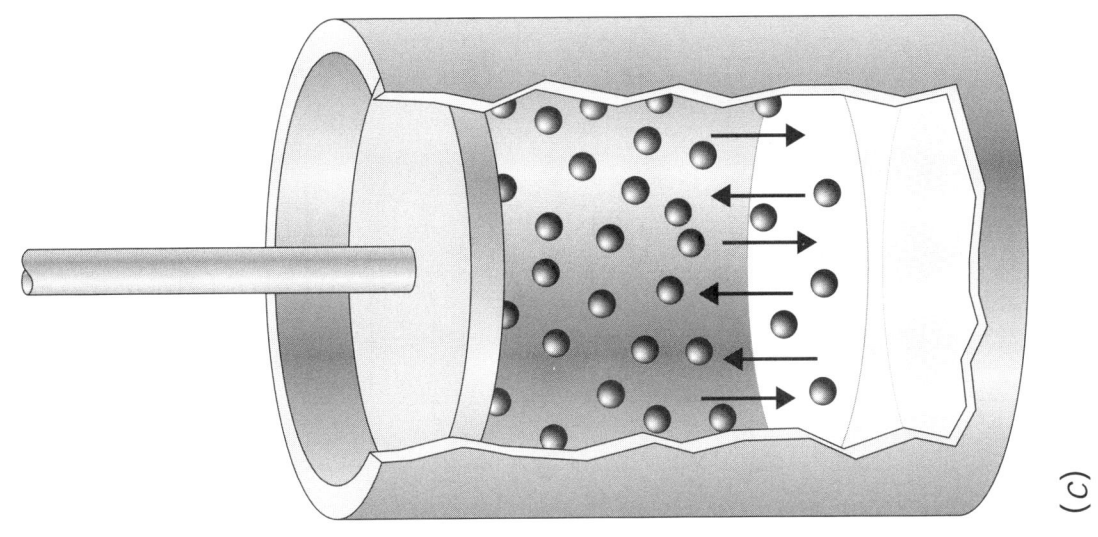

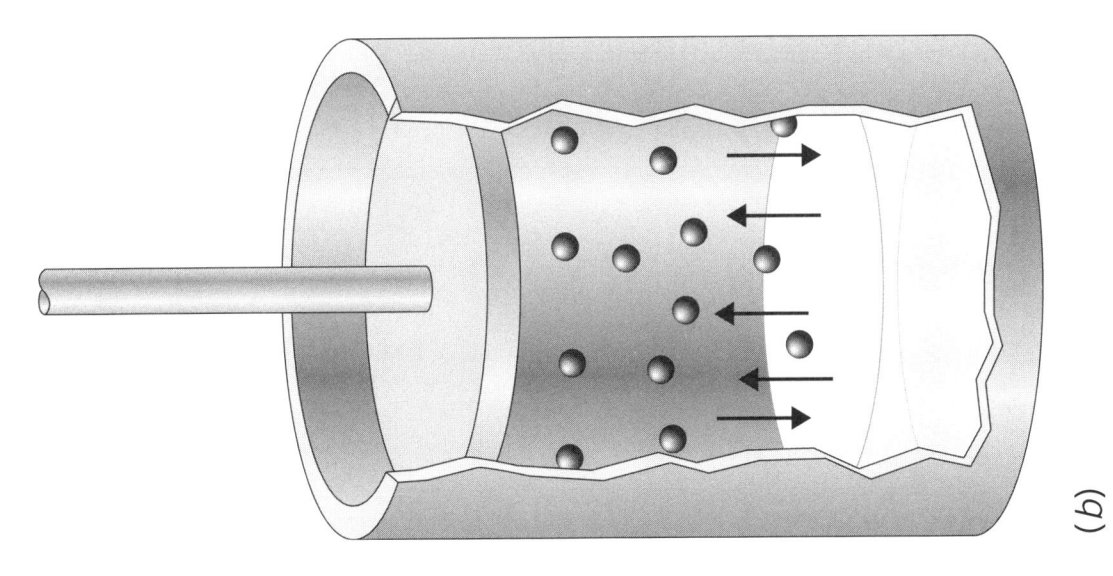

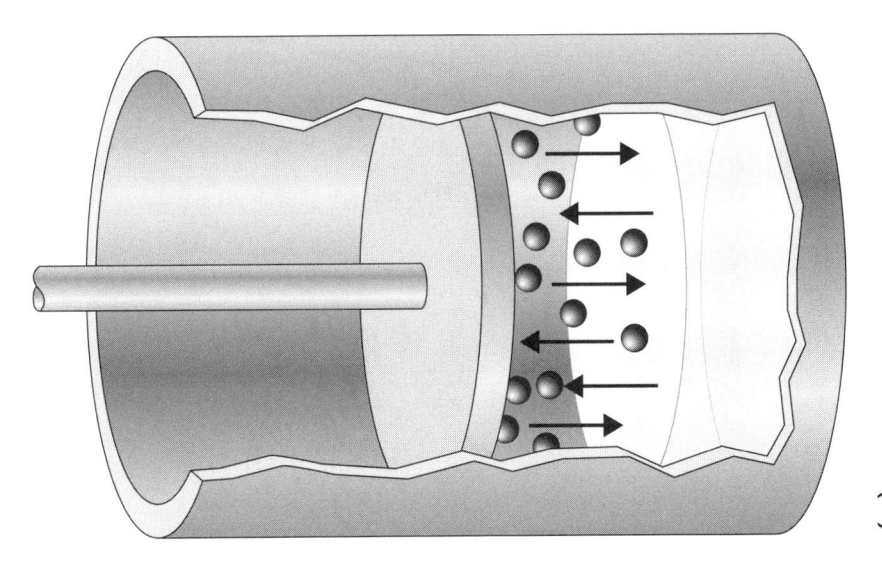

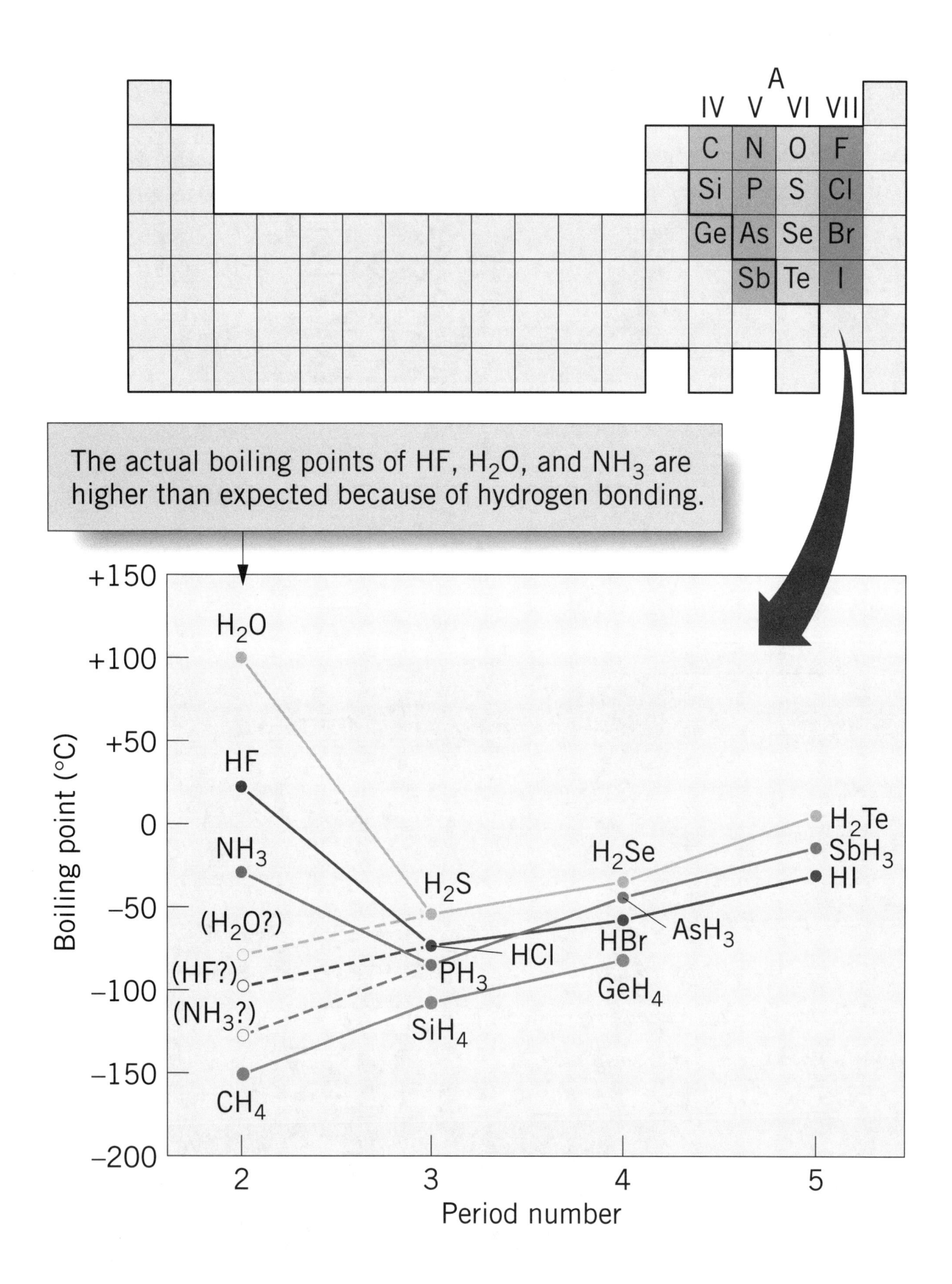

The actual boiling points of HF, H$_2$O, and NH$_3$ are higher than expected because of hydrogen bonding.

FIGURE 11.21

FIGURE 11.24

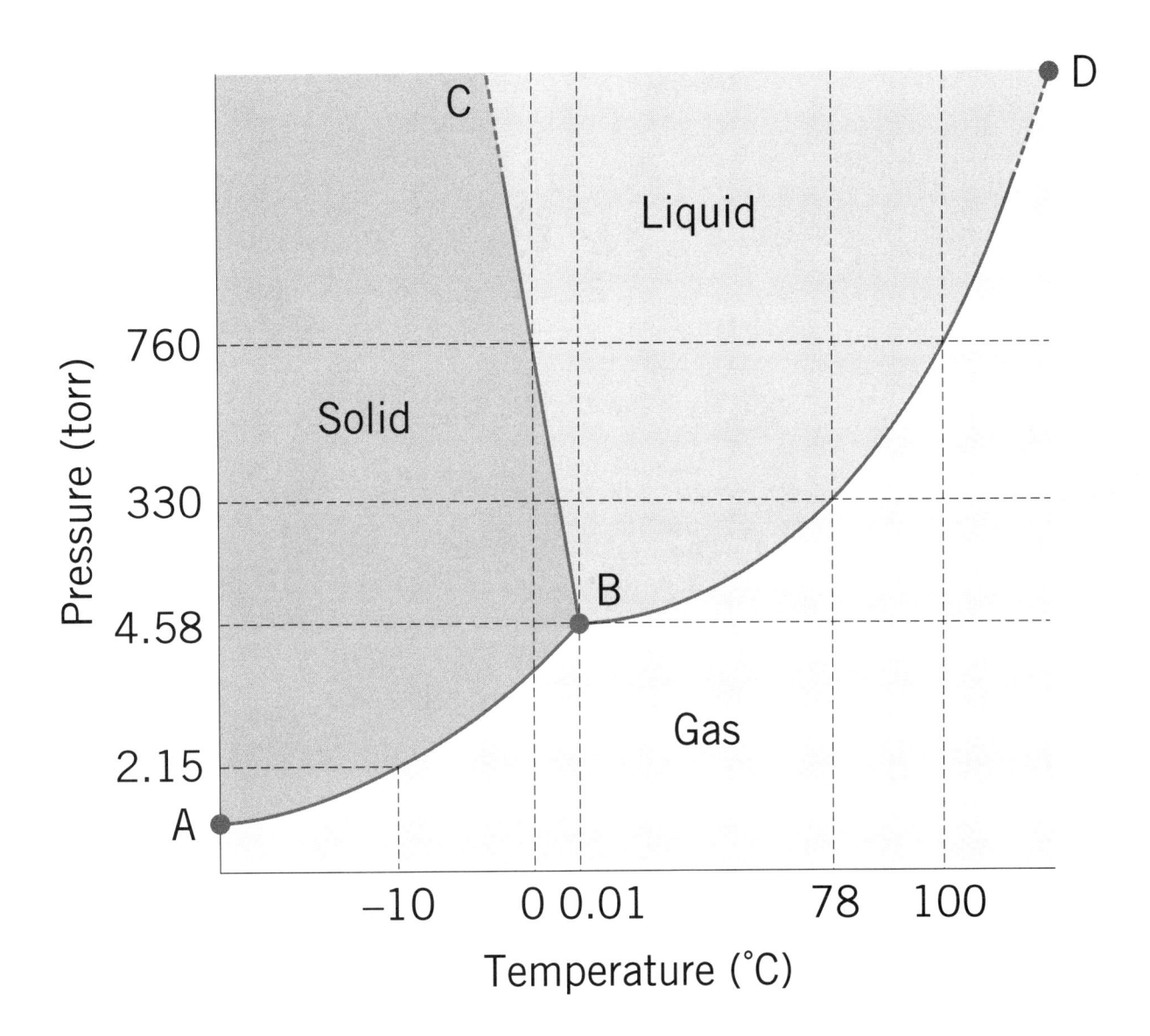

FIGURE 11.25

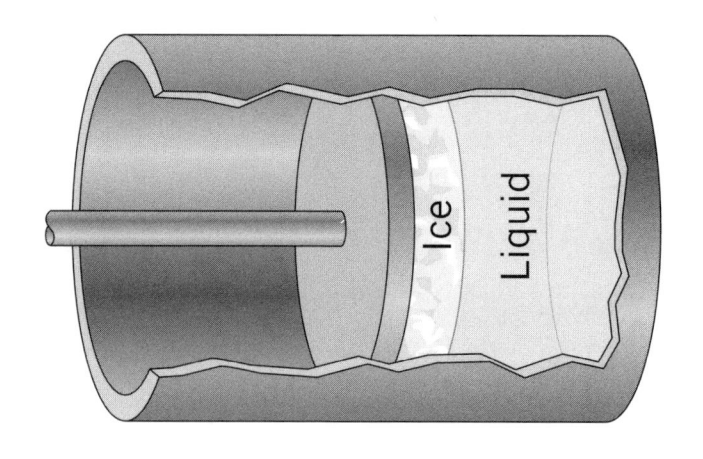

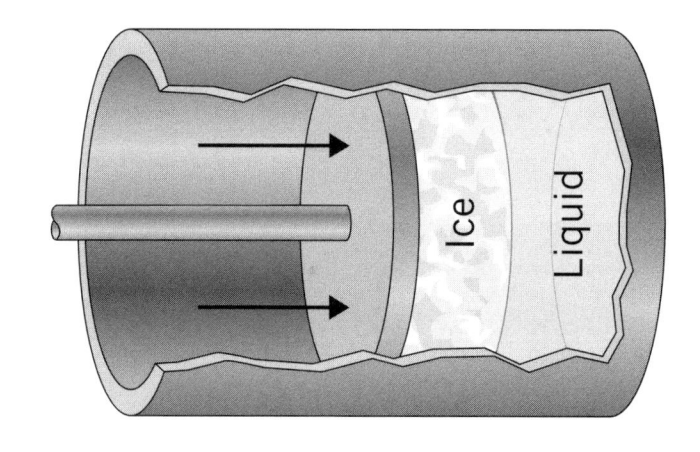

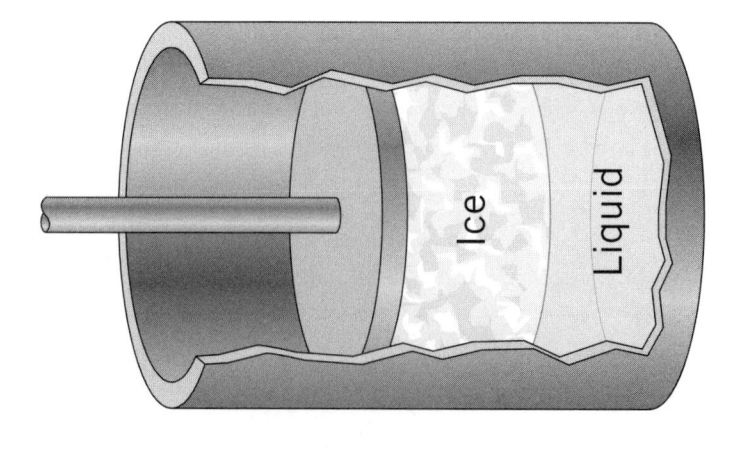

FIGURE 11.29

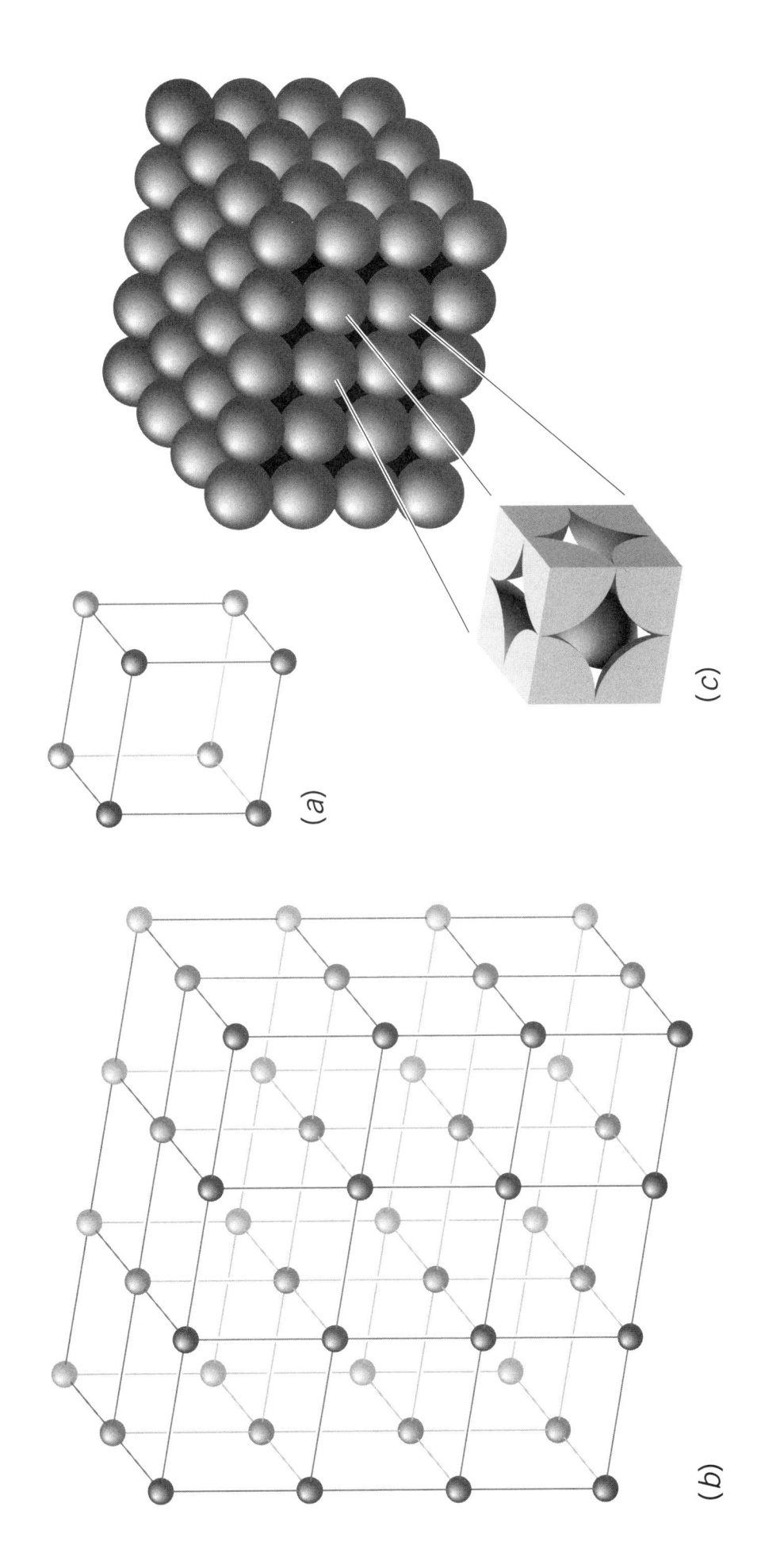

FIGURE 11.30 & 11.32

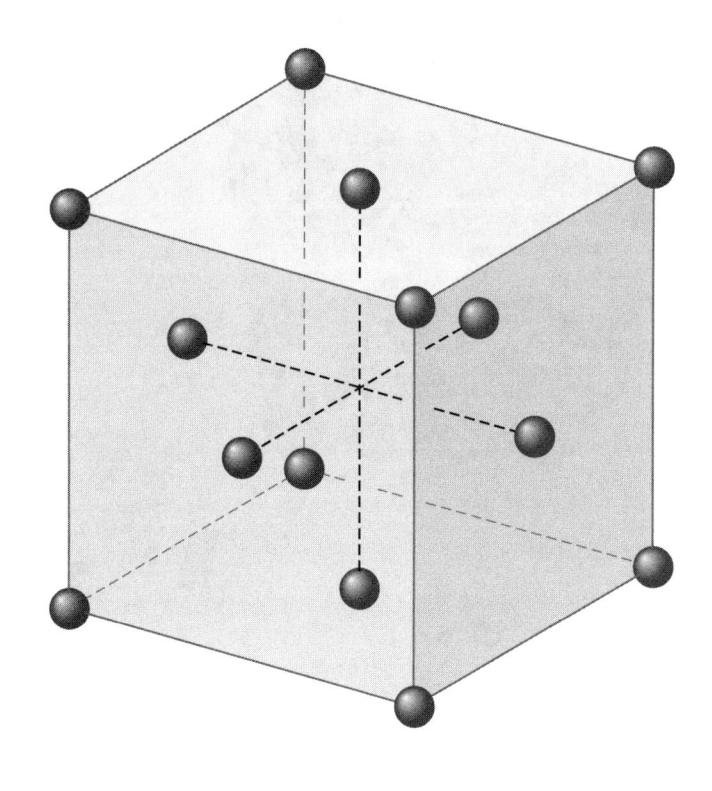

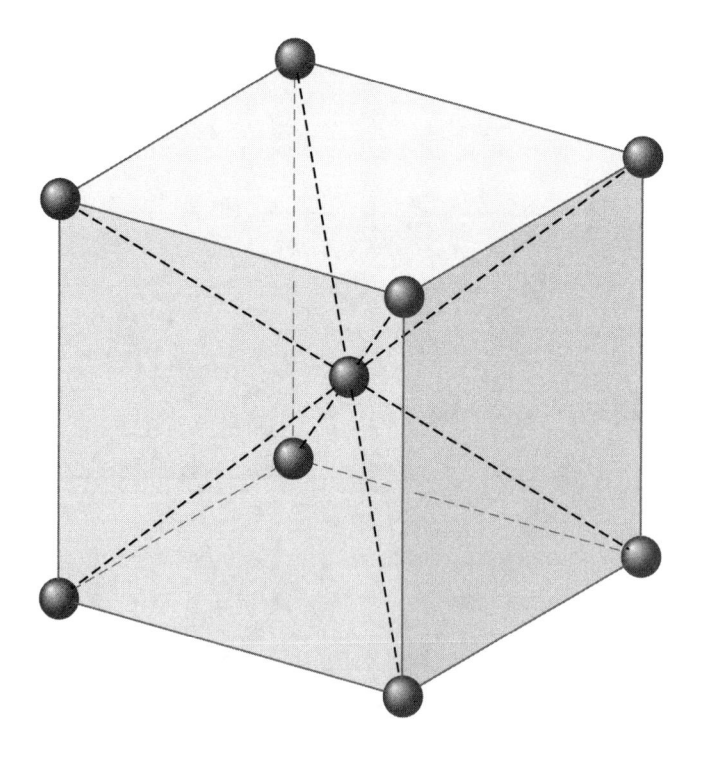

FIGURE 11.33

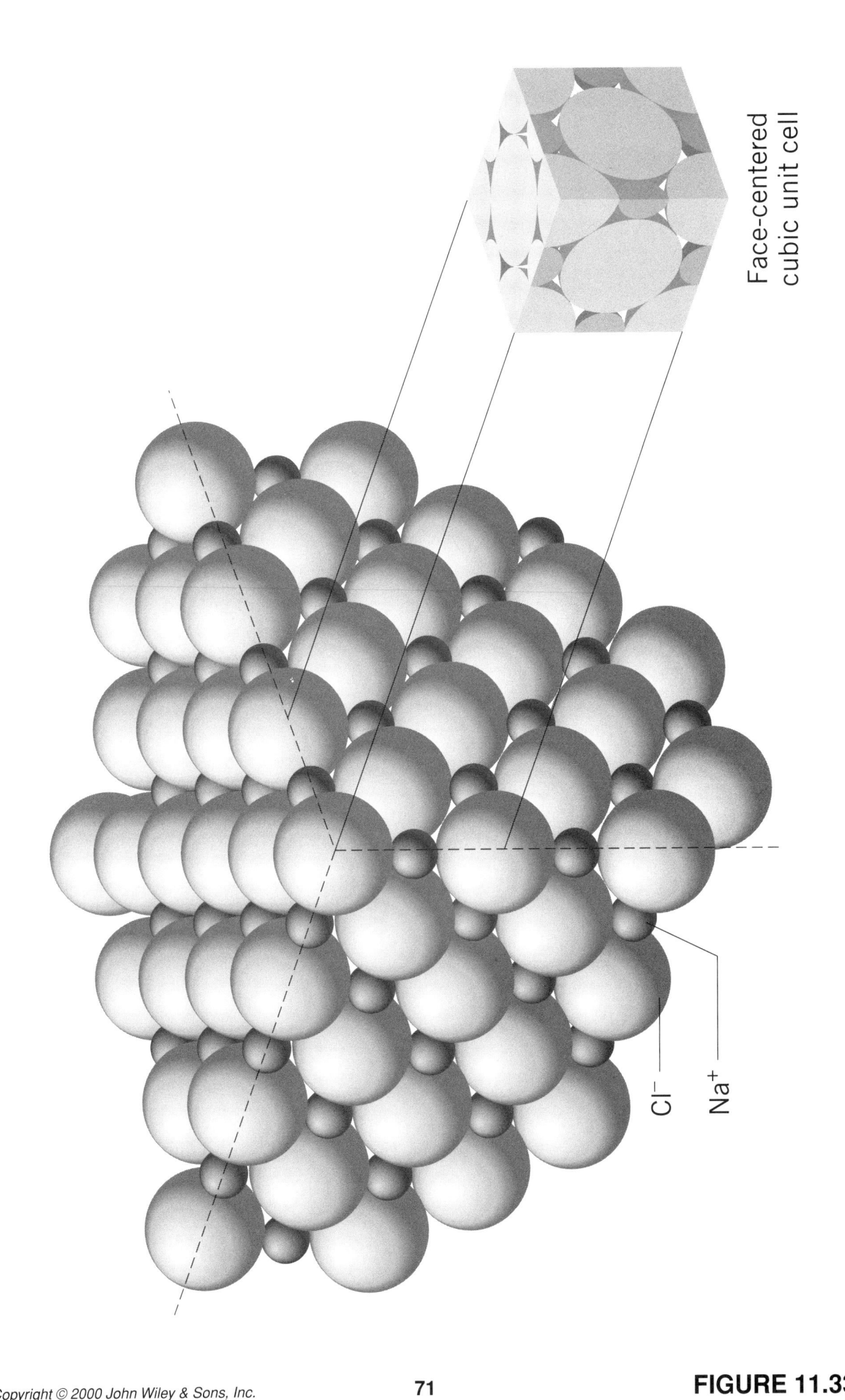

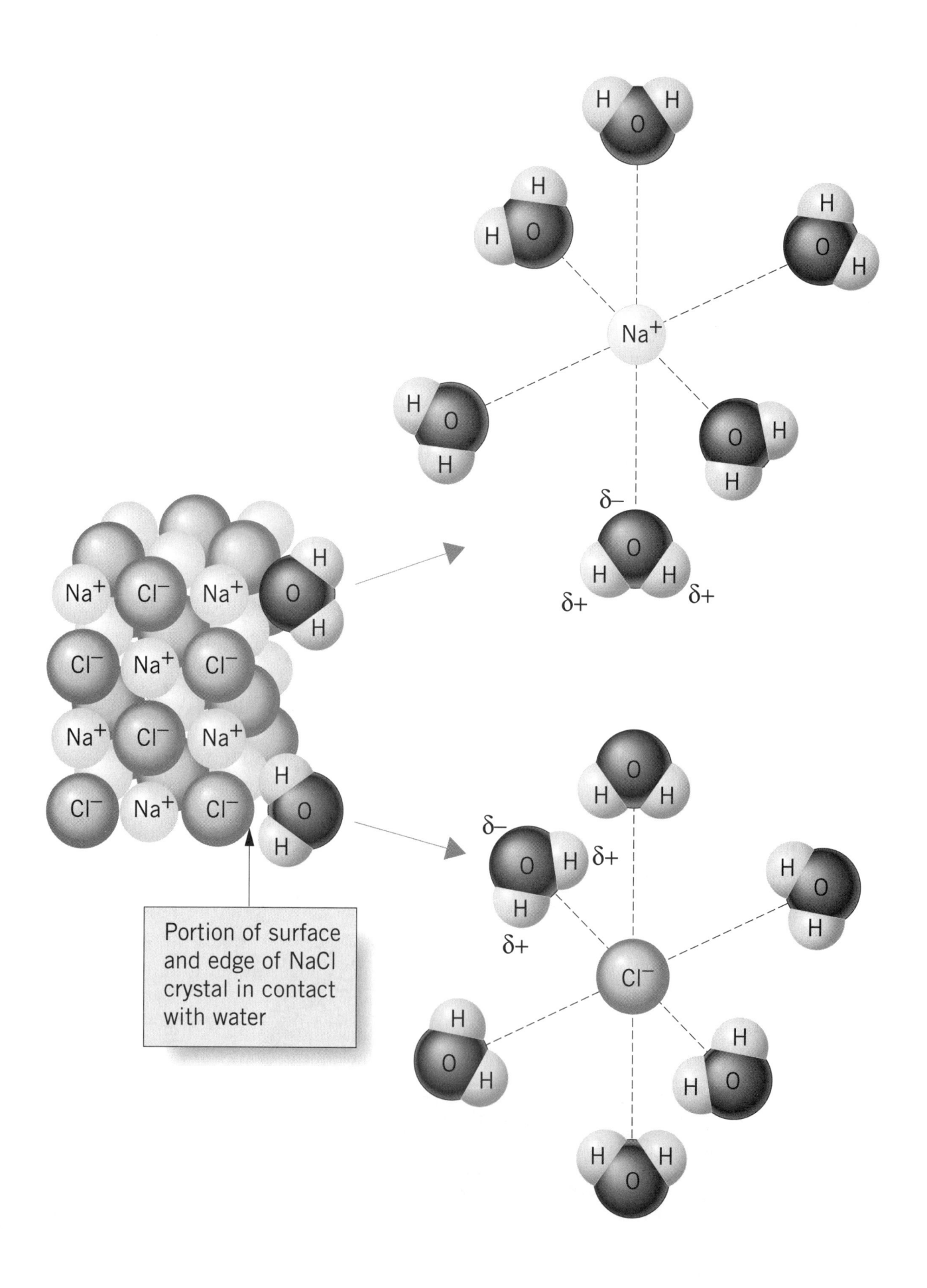

Portion of surface and edge of NaCl crystal in contact with water

FIGURE 12.4

Molecule of polar covalent compound

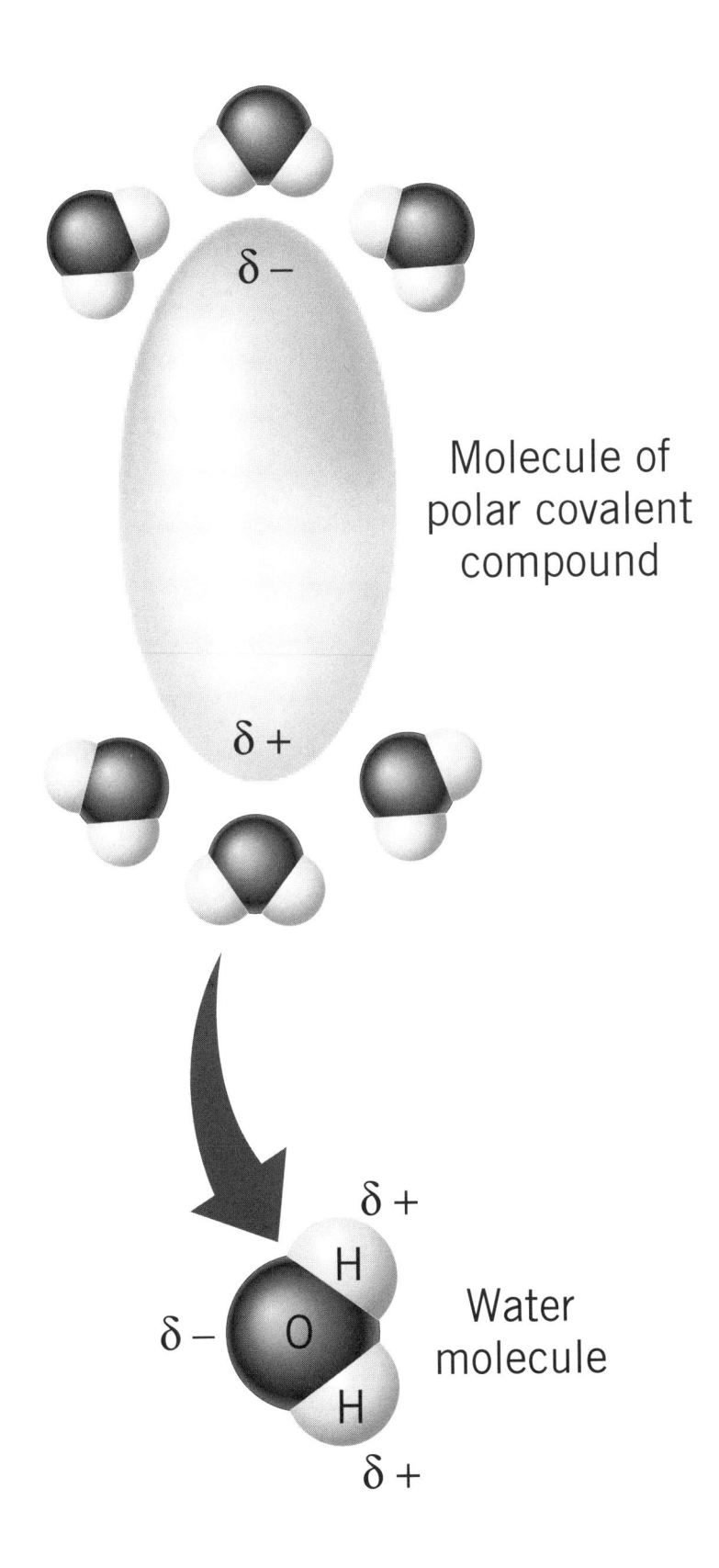

Water molecule

FIGURE 12.5

FIGURE 12.6

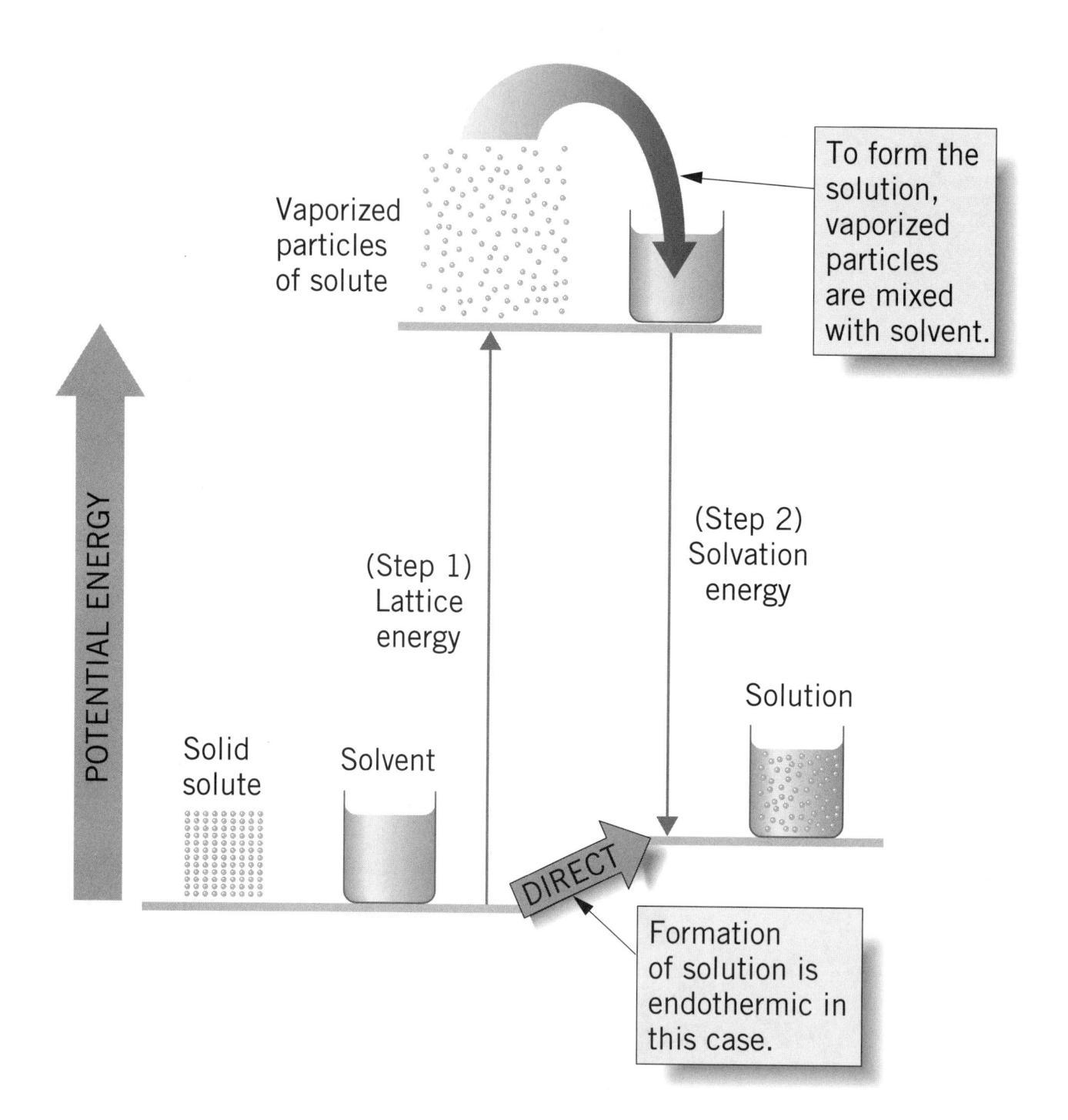

Vaporized particles of solute

To form the solution, vaporized particles are mixed with solvent.

POTENTIAL ENERGY

(Step 1) Lattice energy

(Step 2) Solvation energy

Solid solute

Solvent

Solution

DIRECT

Formation of solution is endothermic in this case.

FIGURE 12.9

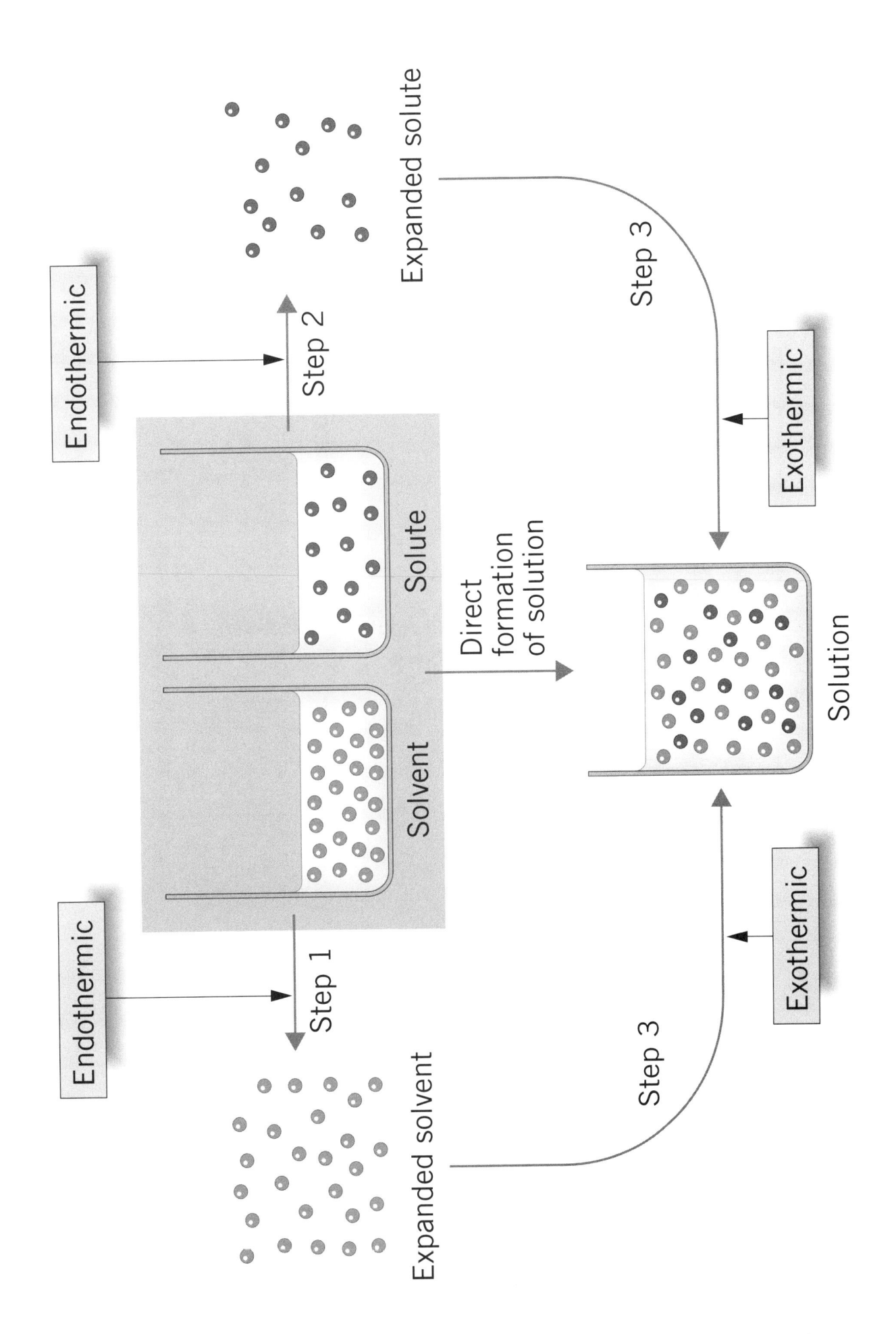

FIGURE 12.10

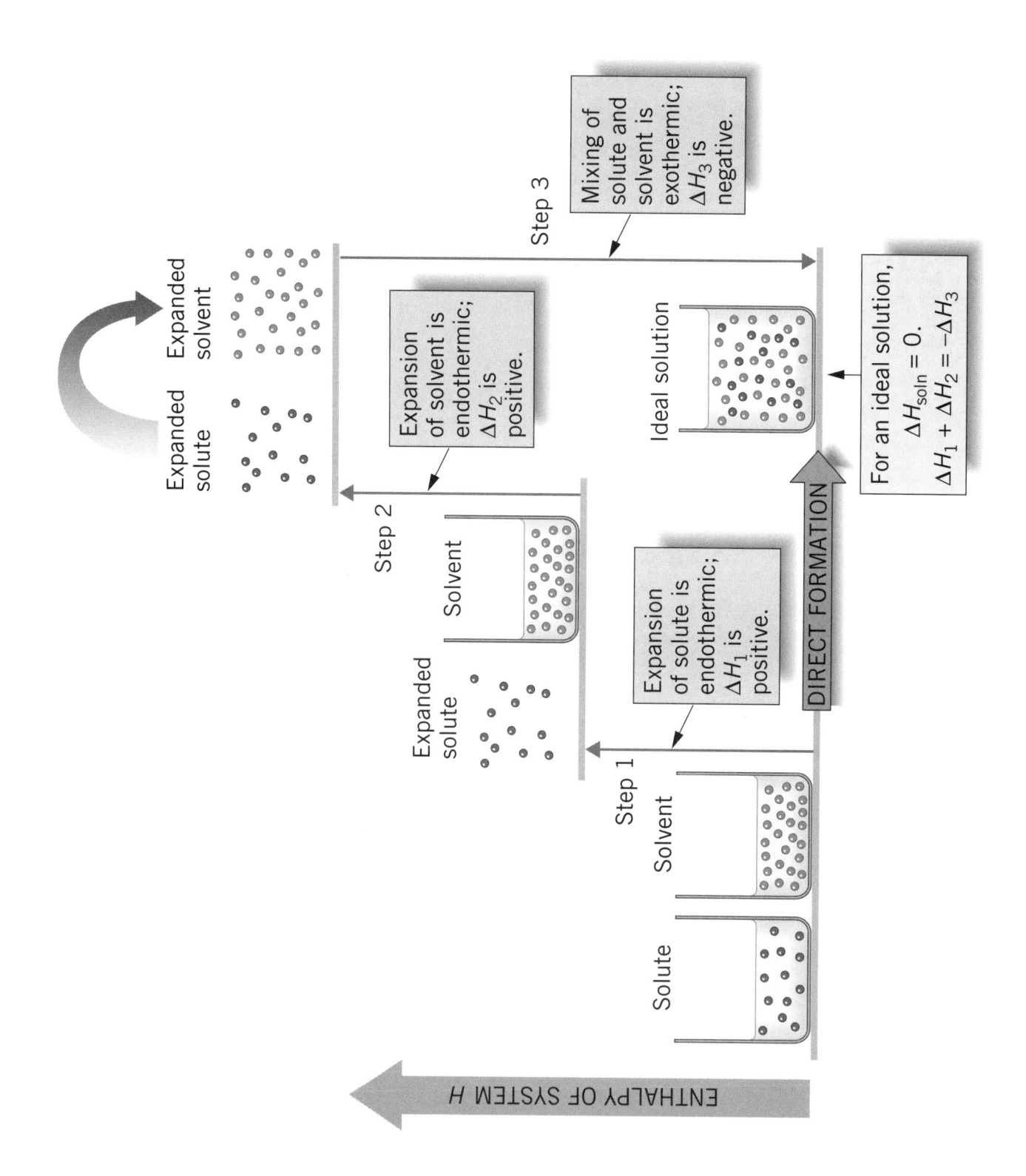

FIGURE 12.11

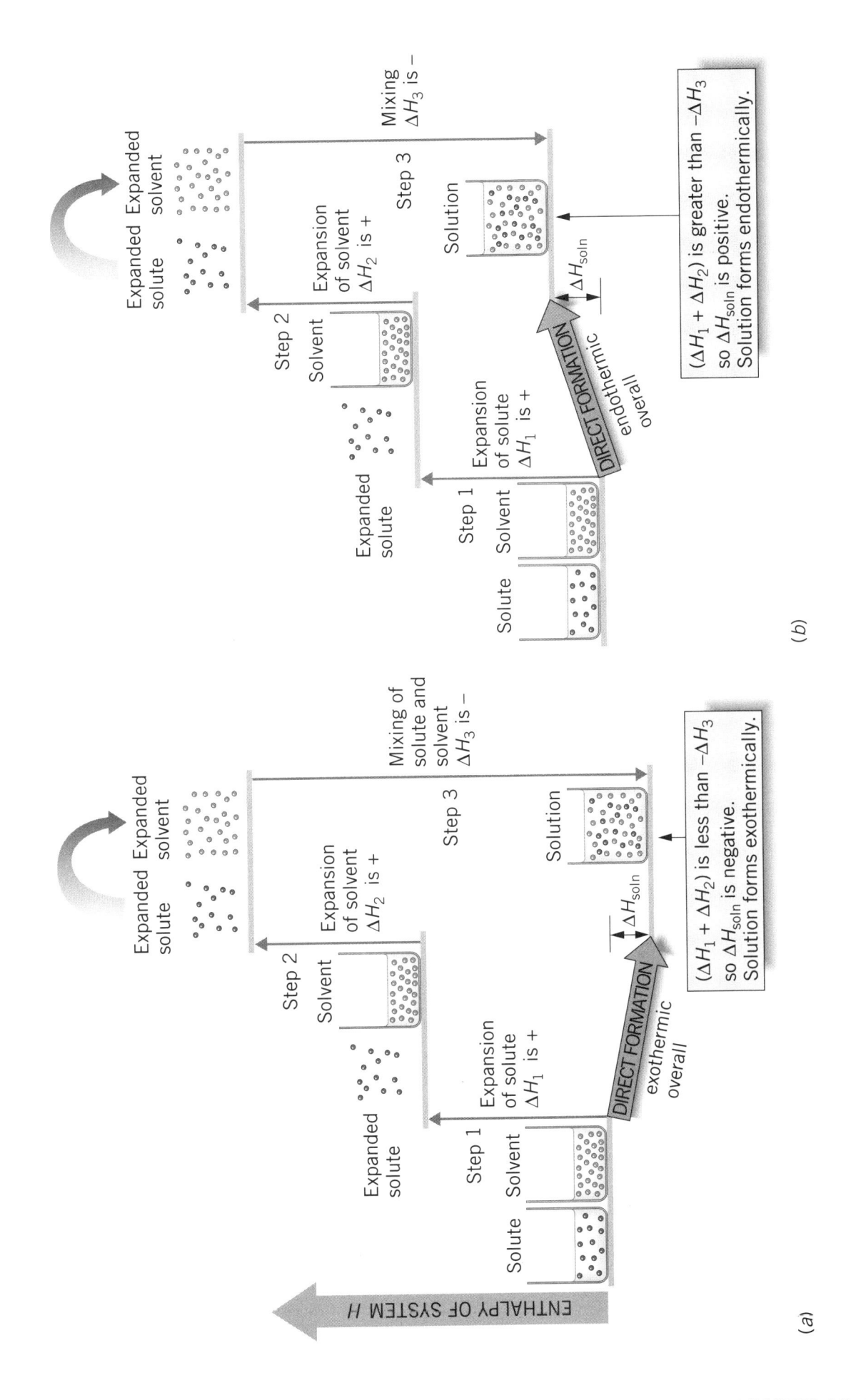

FIGURE 12.13

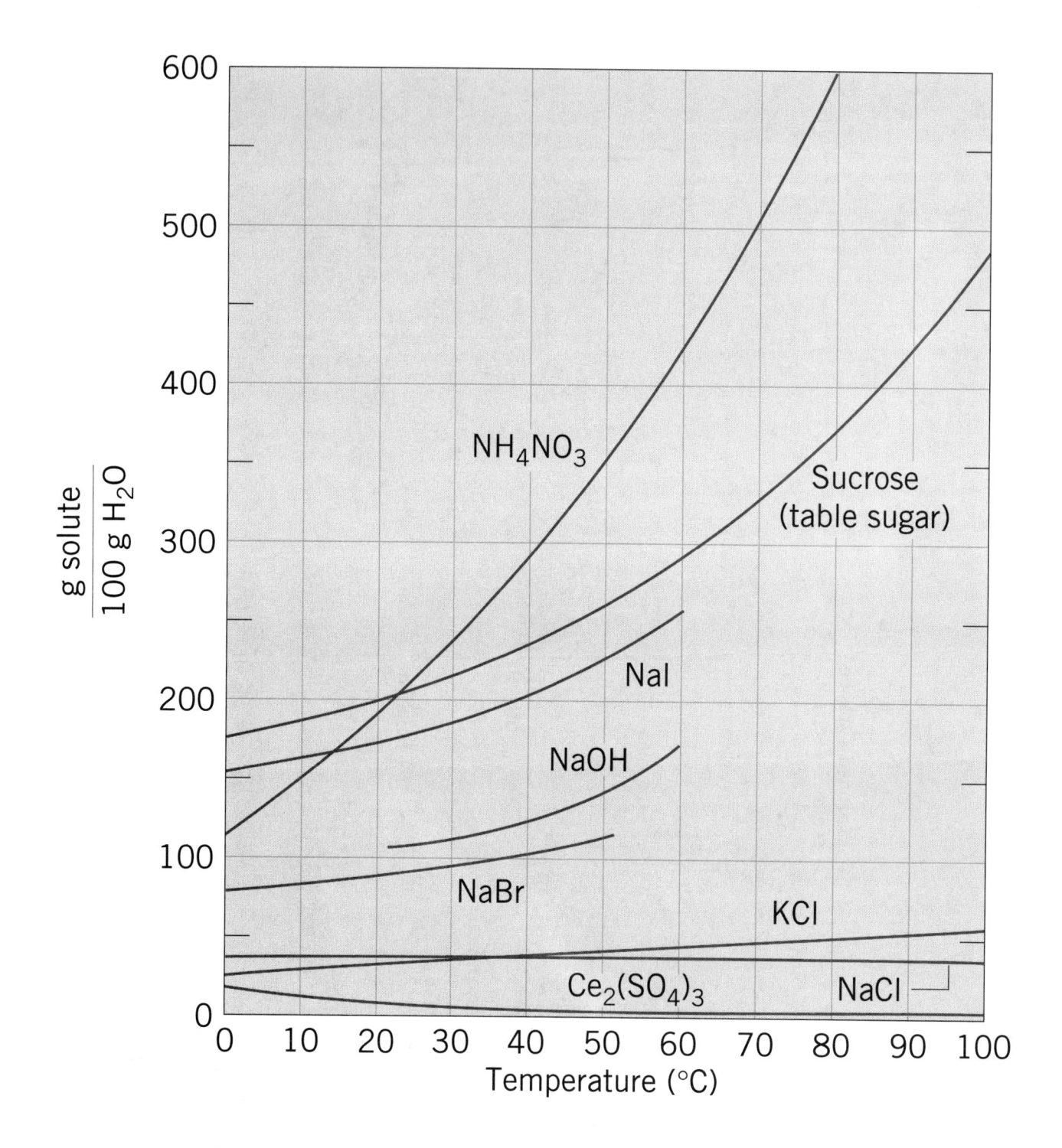

FIGURE 12.15

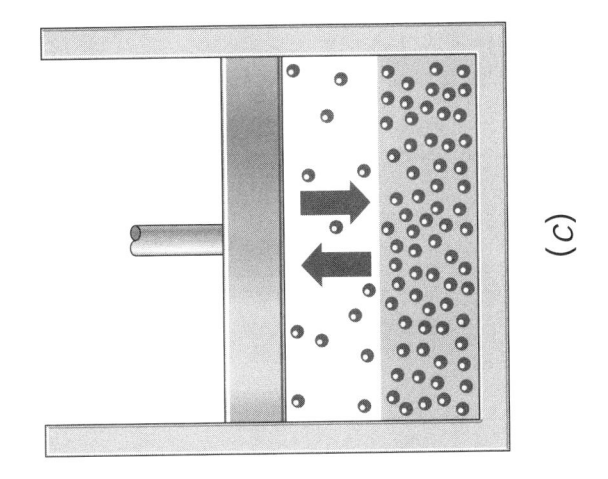

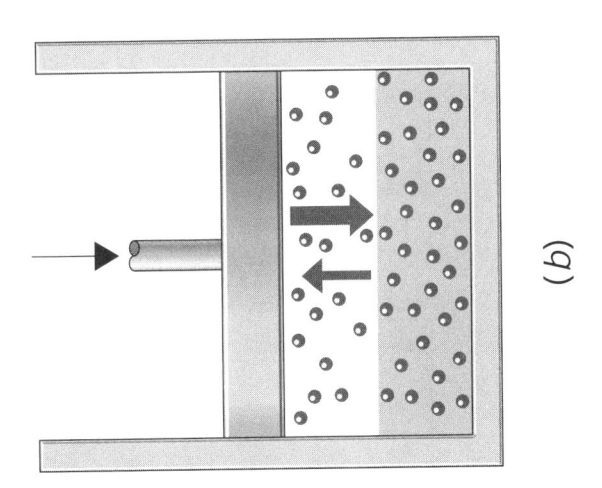

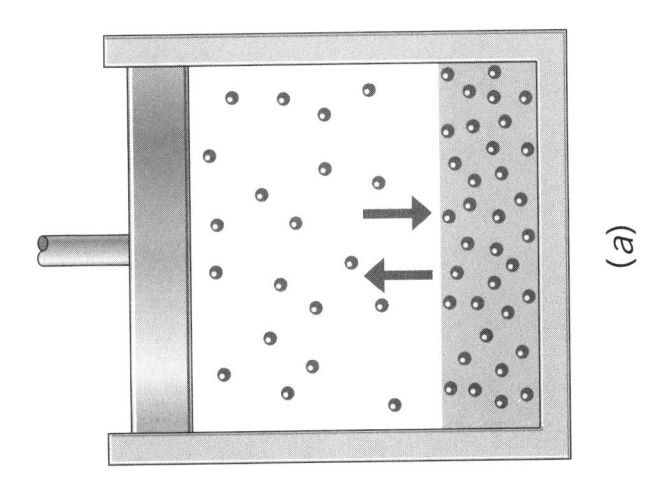

Vapor pressure gauges

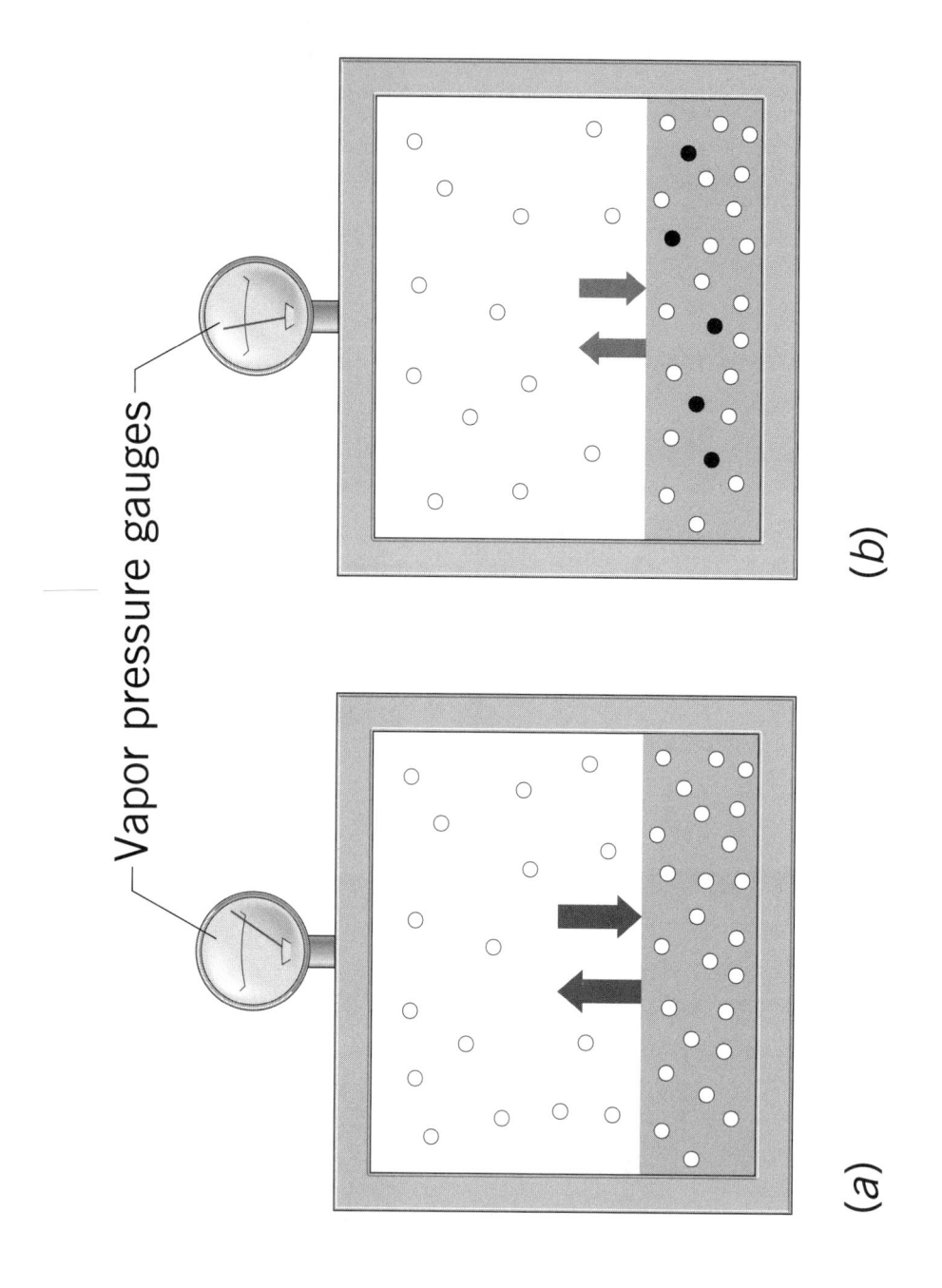

(a)

(b)

FIGURE 12.17

FIGURE 12.20

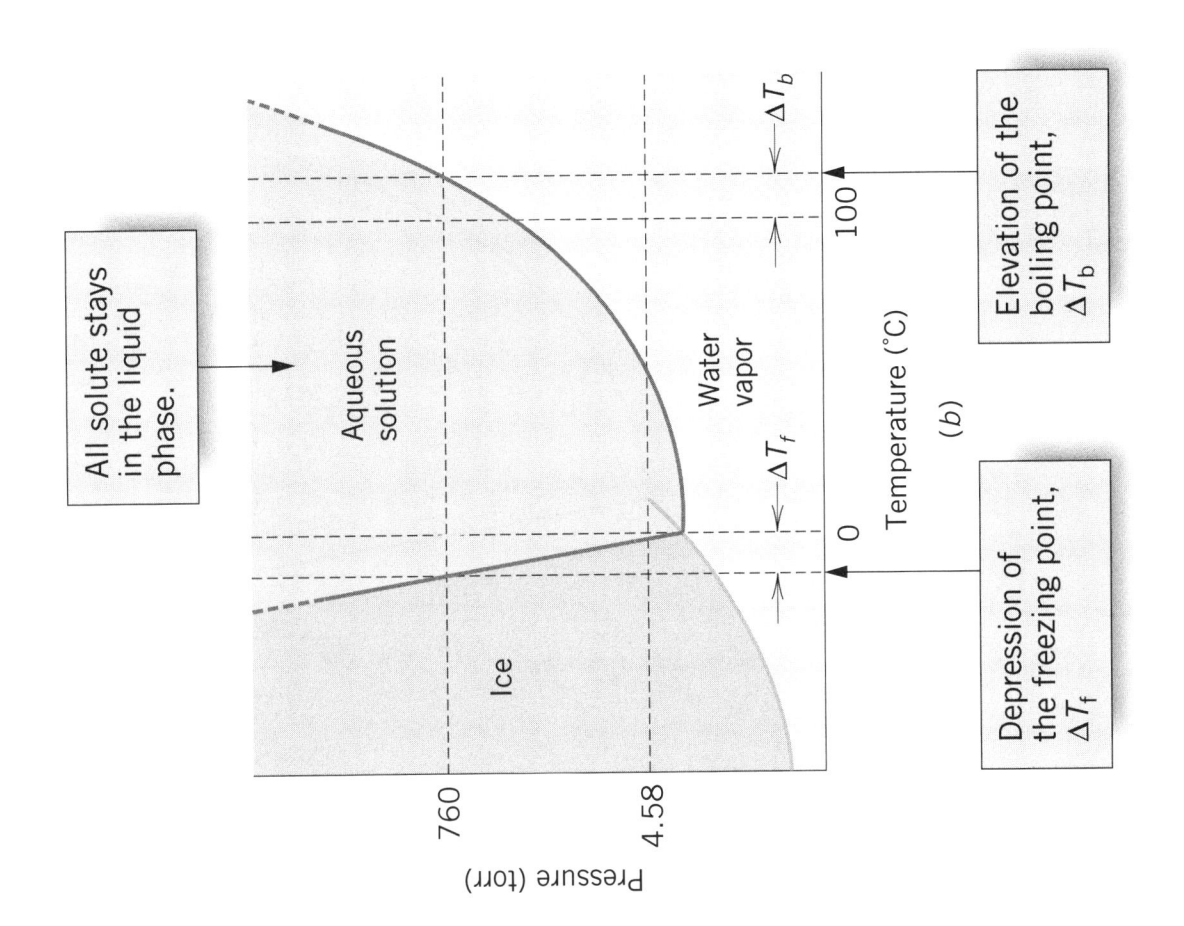

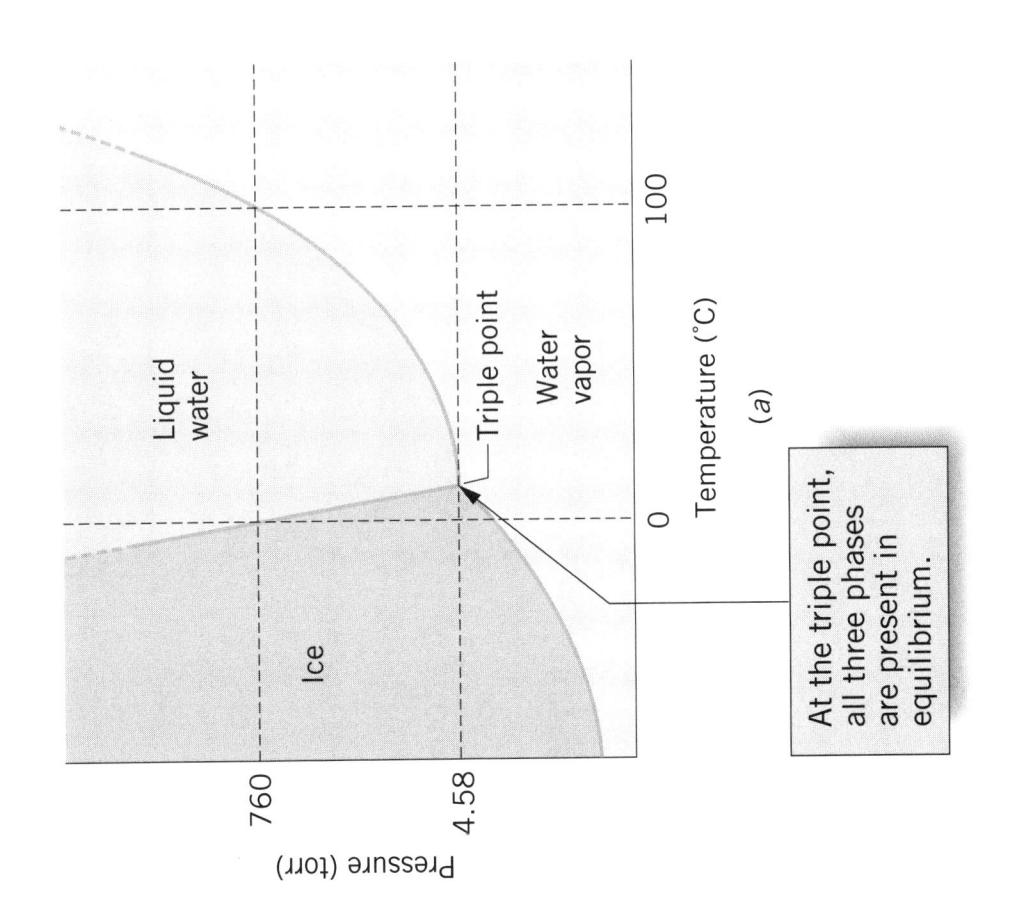

Solution
Osmotic membrane
Pure water

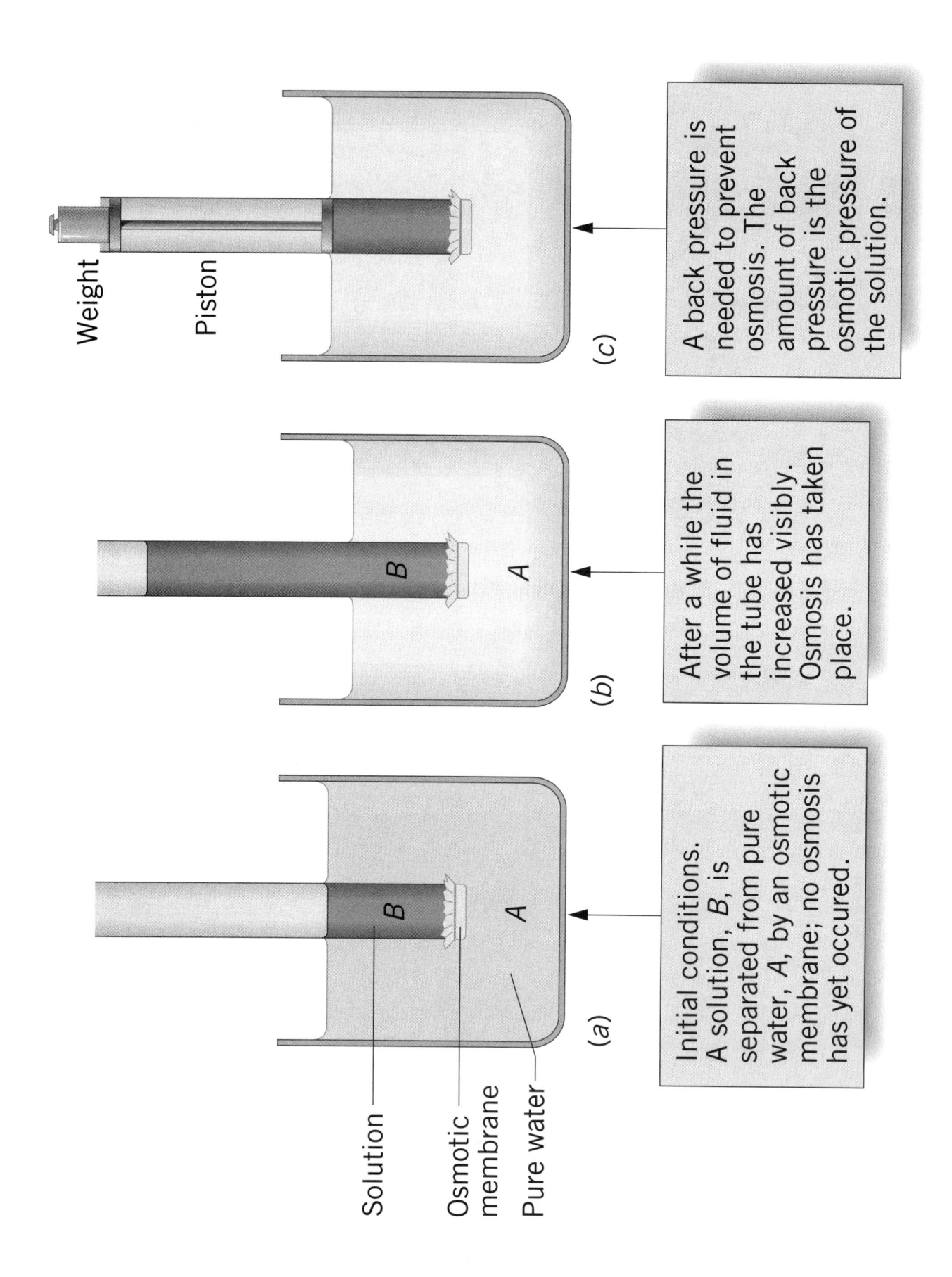

(a) Initial conditions. A solution, B, is separated from pure water, A, by an osmotic membrane; no osmosis has yet occured.

(b) After a while the volume of fluid in the tube has increased visibly. Osmosis has taken place.

Weight
Piston

(c) A back pressure is needed to prevent osmosis. The amount of back pressure is the osmotic pressure of the solution.

FIGURE 12.21

At this instant in time

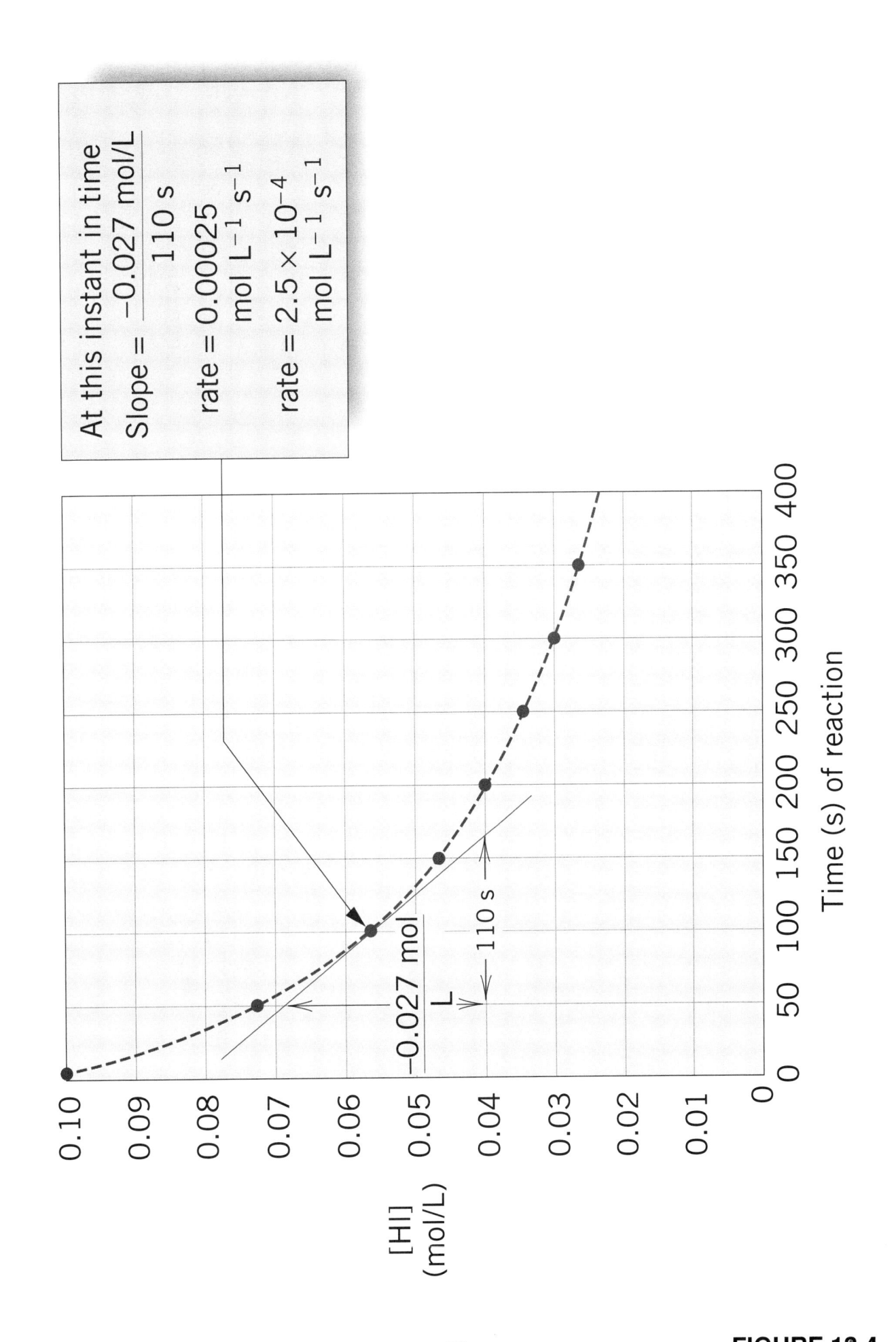

Slope = $\dfrac{-0.027 \text{ mol/L}}{110 \text{ s}}$

rate = 0.00025 mol L^{-1} s^{-1}

rate = 2.5×10^{-4} mol L^{-1} s^{-1}

[HI] (mol/L) vs. Time (s) of reaction

−0.027 mol/L / 110 s

FIGURE 13.4

FIGURE 13.5

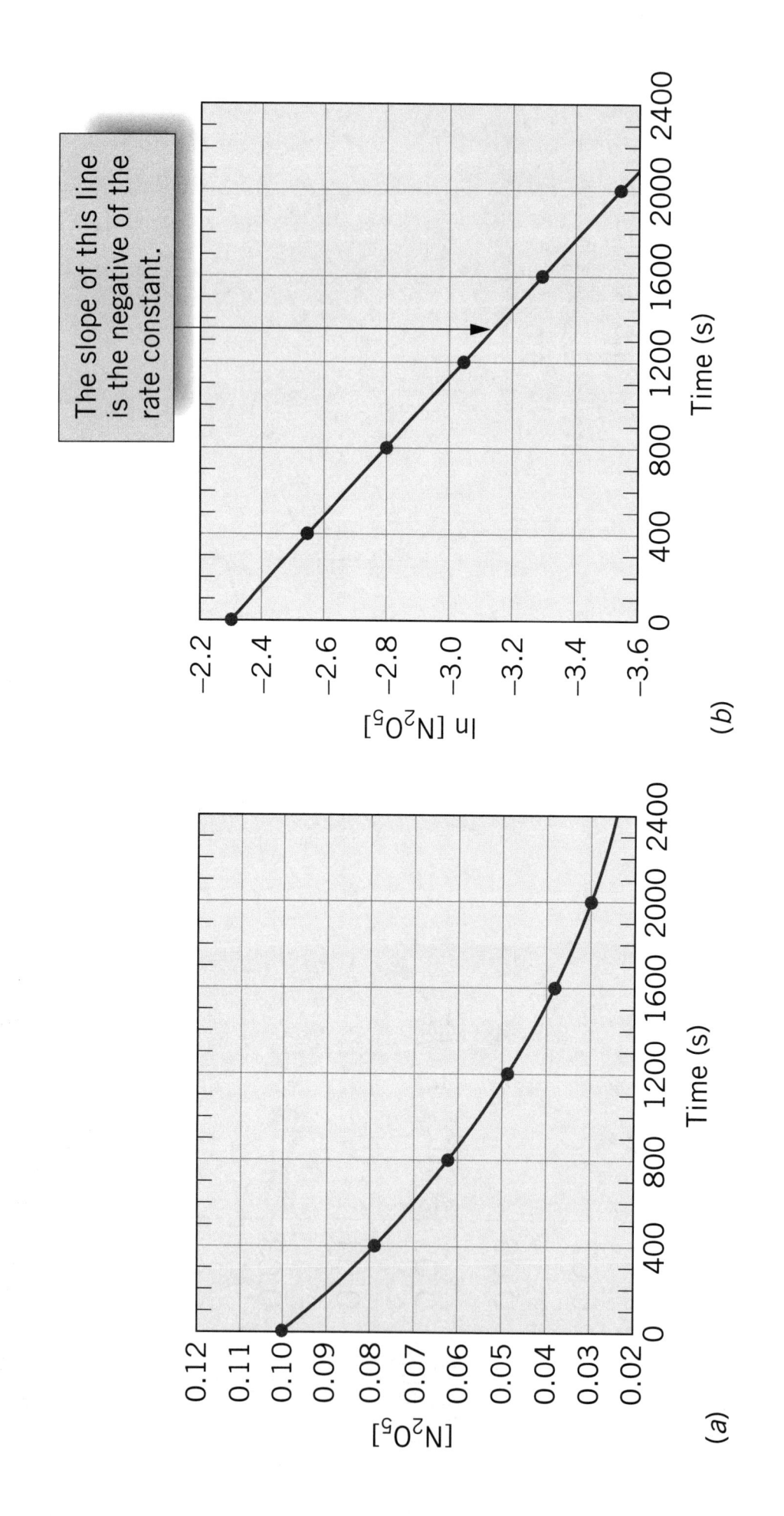

FIGURE 13.7

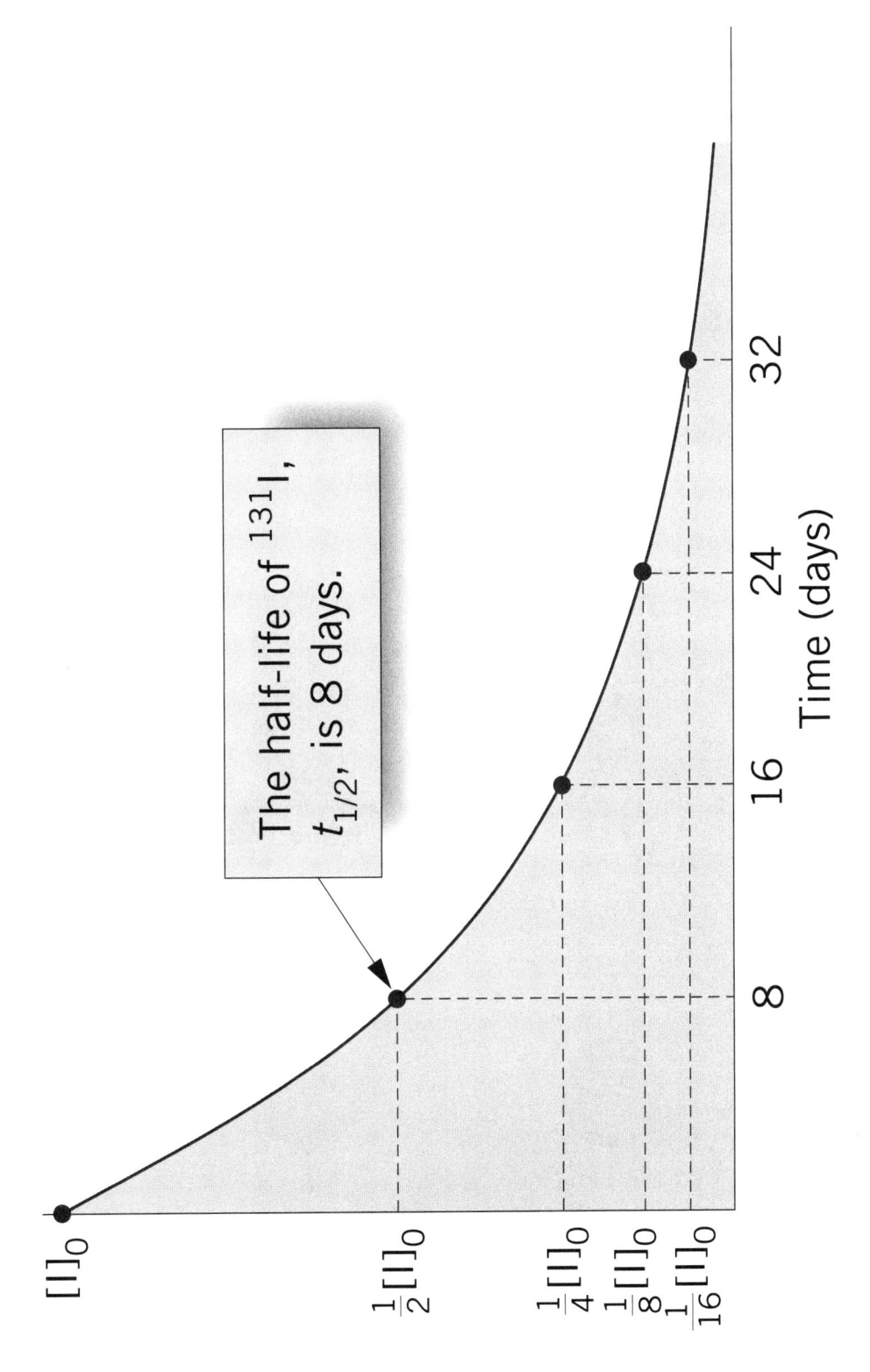

FIGURE 13.8

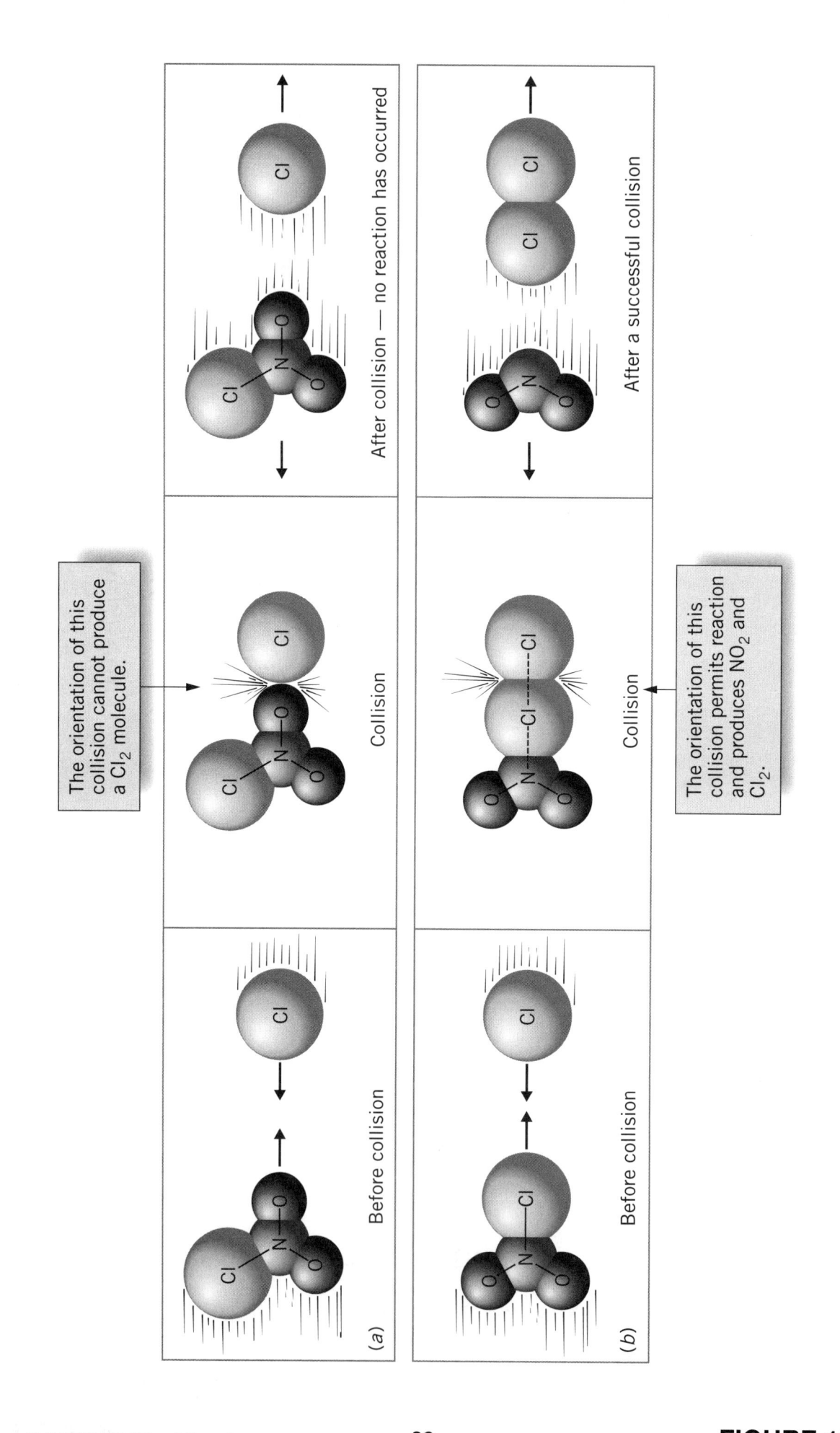

FIGURE 13.14

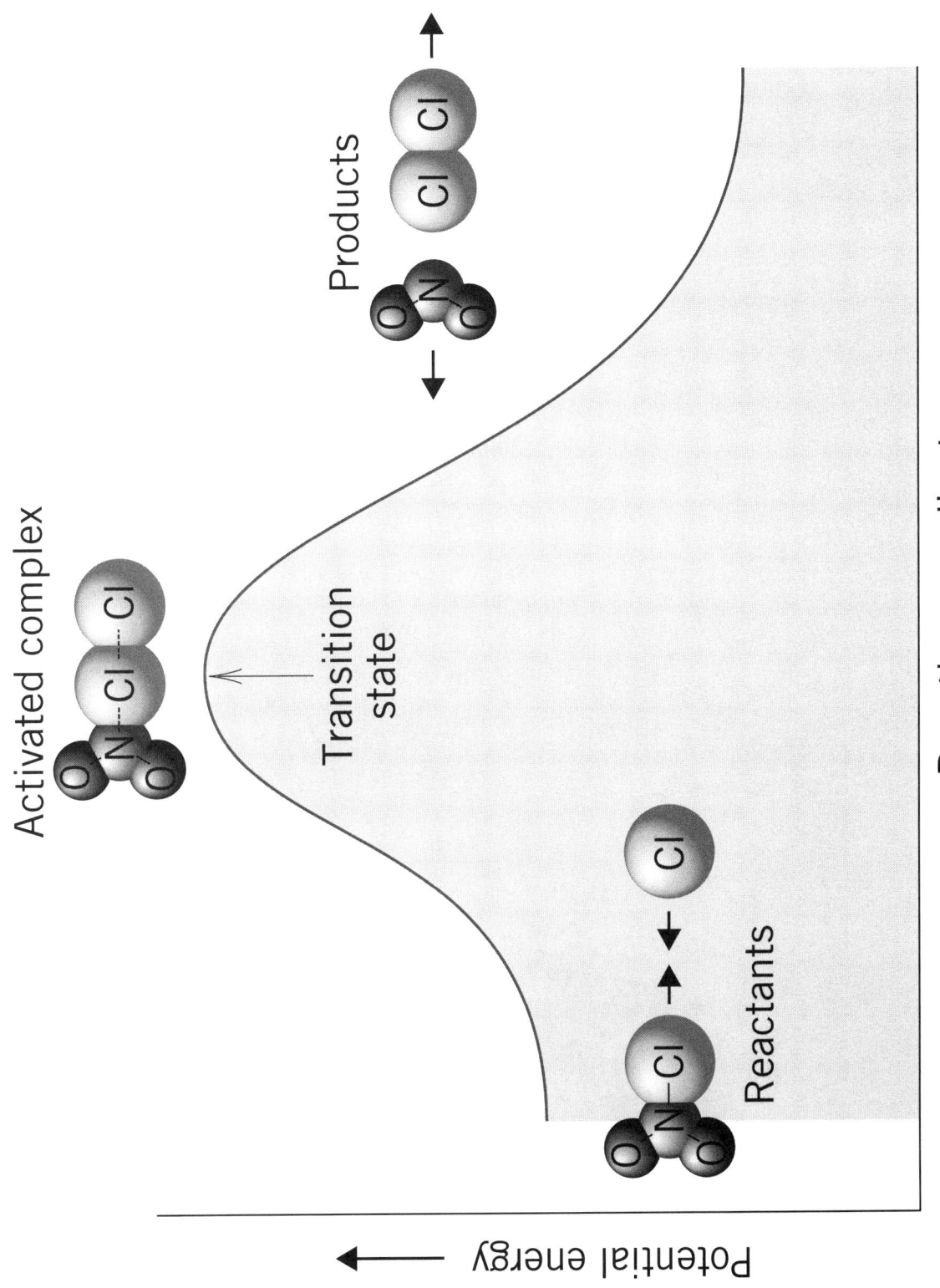

FIGURE 13.15

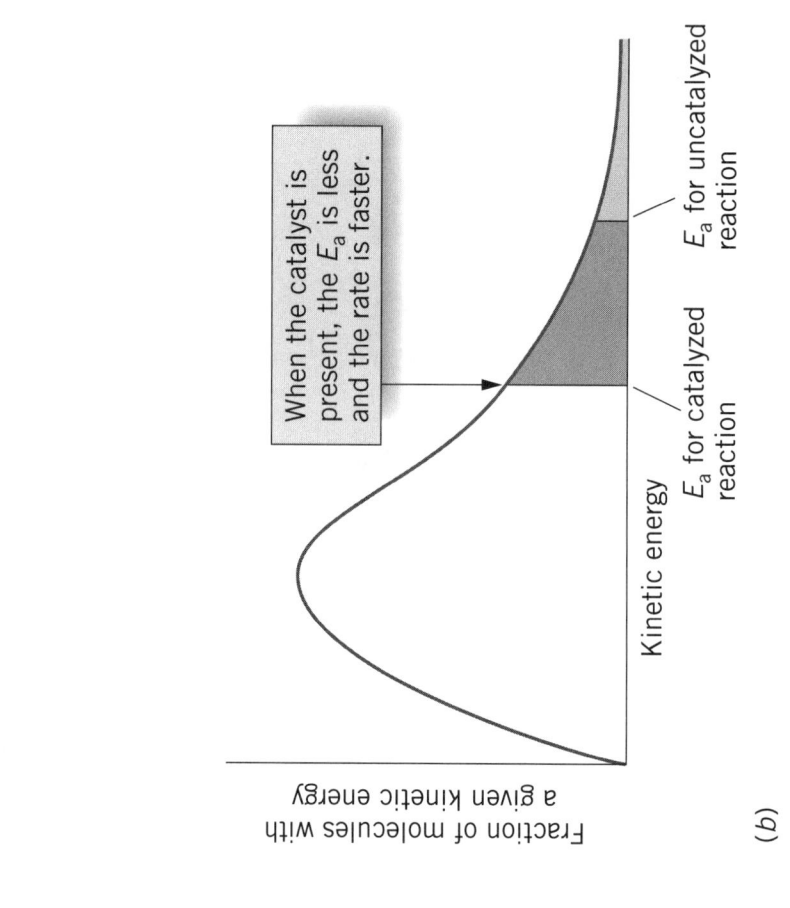

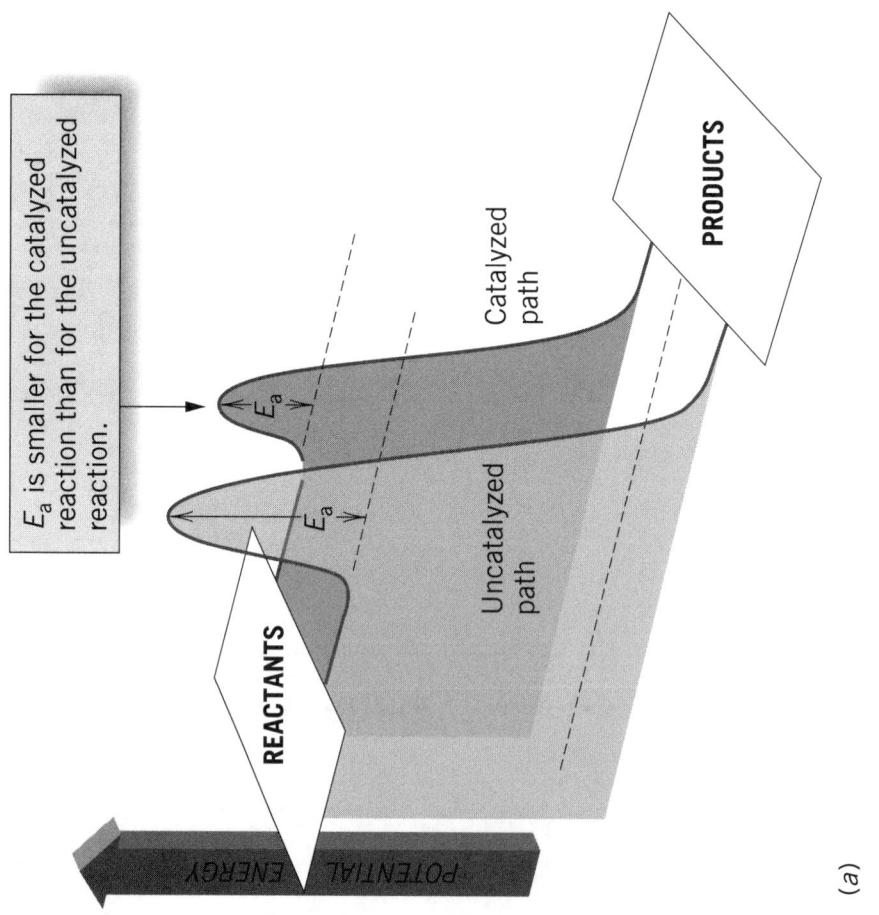

FIGURE 13.16

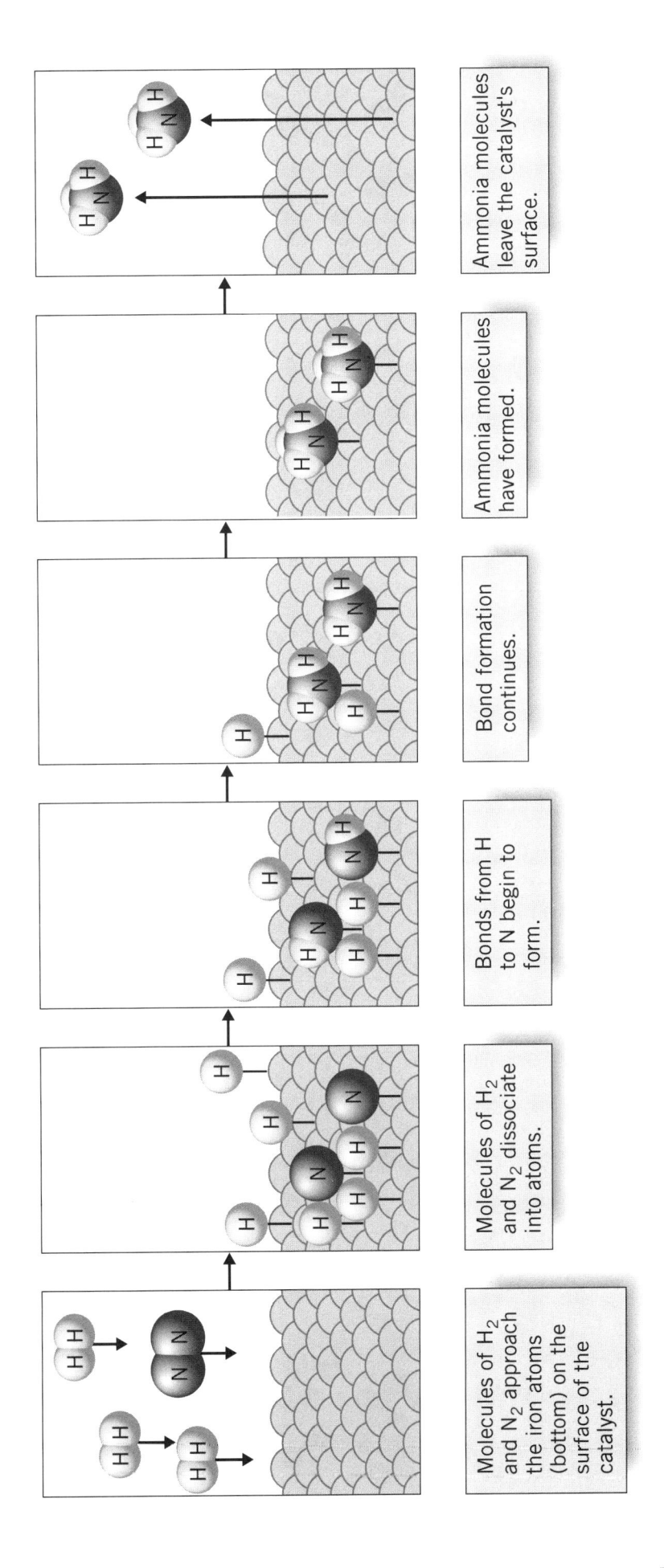

1 liter

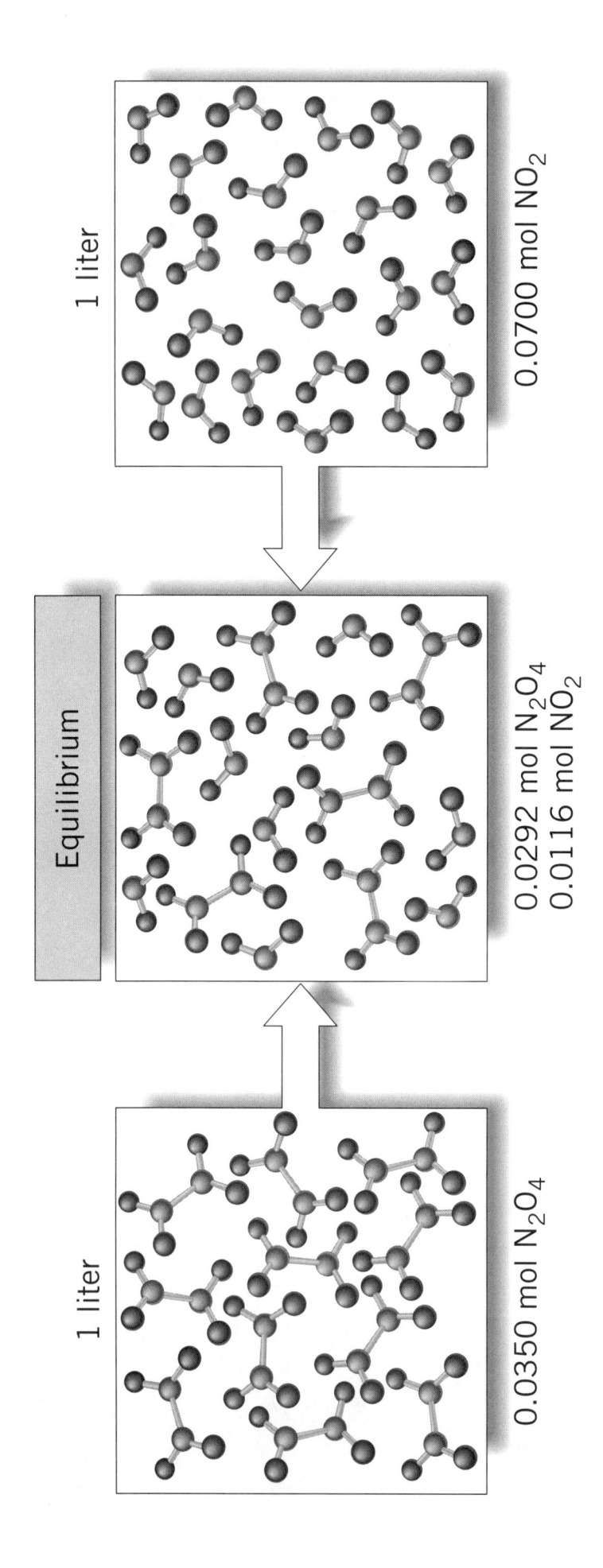

0.0700 mol NO$_2$

Equilibrium

0.0292 mol N$_2$O$_4$
0.0116 mol NO$_2$

1 liter

0.0350 mol N$_2$O$_4$

FIGURE 14.2

HCHO$_2$ (acid) + H$_2$O (base) → CHO$_2^-$ + H$_3$O$^+$

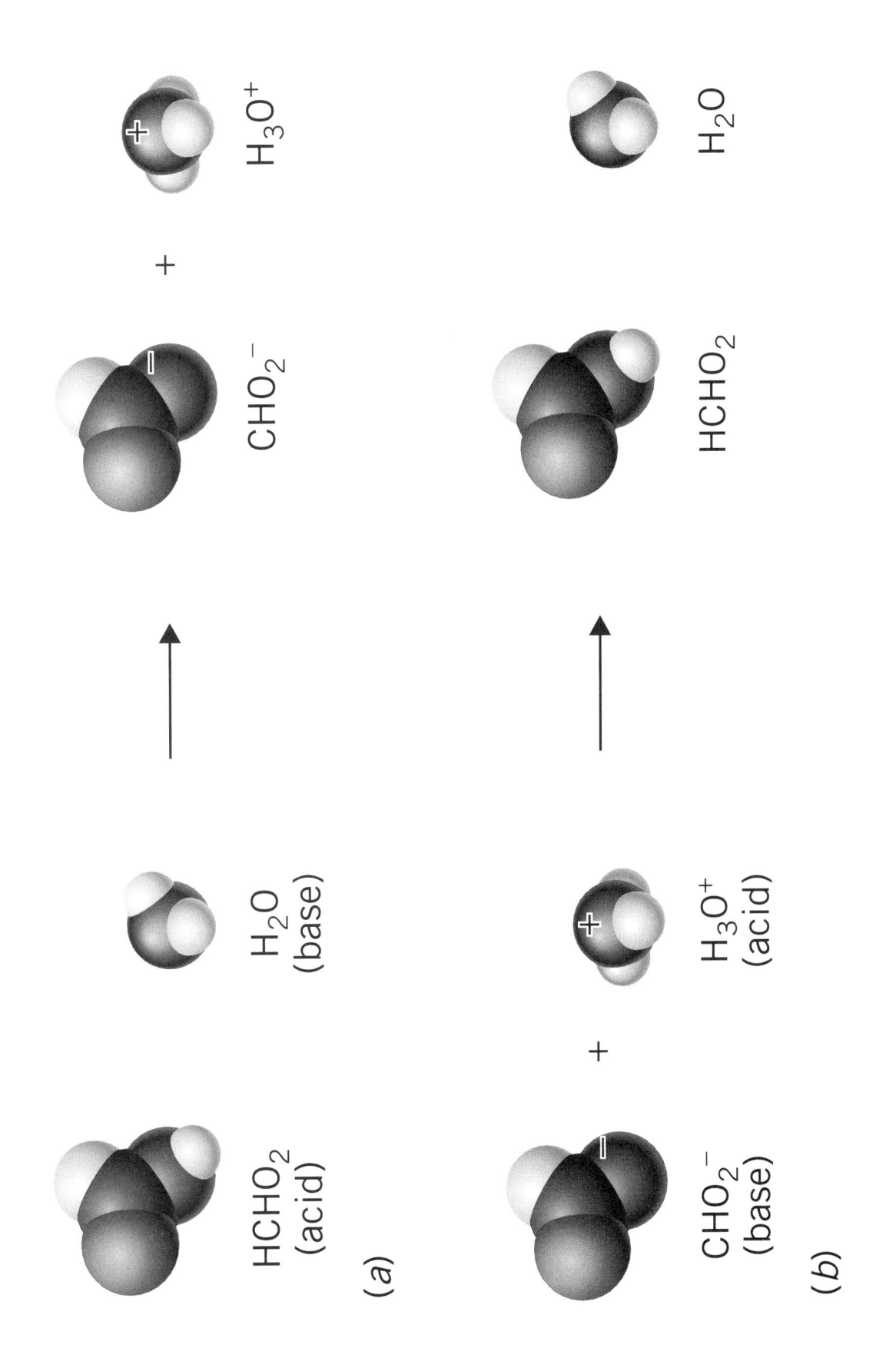

(a)

CHO$_2^-$ (base) + H$_3$O$^+$ (acid) → HCHO$_2$ + H$_2$O

(b)

FIGURE 15.2

pH

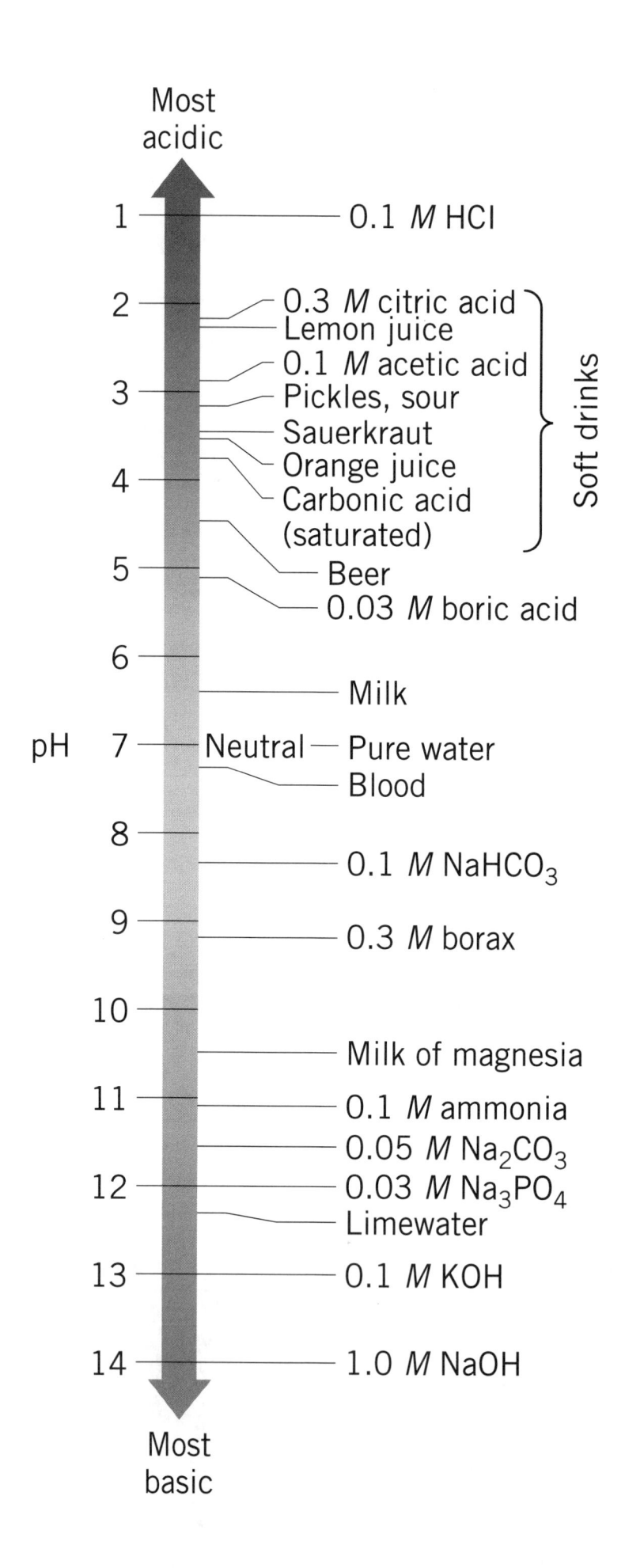

- 1 — 0.1 M HCl
- 2 — 0.3 M citric acid
- Lemon juice
- 0.1 M acetic acid
- 3 — Pickles, sour
- Sauerkraut
- Orange juice
- 4 — Carbonic acid (saturated)
- Beer
- 5 — 0.03 M boric acid
- 6
- Milk
- 7 — Neutral — Pure water
- Blood
- 8
- 0.1 M NaHCO₃
- 9 — 0.3 M borax
- 10
- Milk of magnesia
- 11 — 0.1 M ammonia
- 0.05 M Na₂CO₃
- 12 — 0.03 M Na₃PO₄
- Limewater
- 13 — 0.1 M KOH
- 14 — 1.0 M NaOH

Most acidic ↑

Most basic ↓

Soft drinks: (0.3 M citric acid through Carbonic acid saturated)

FIGURE 15.3

FIGURE 16.5

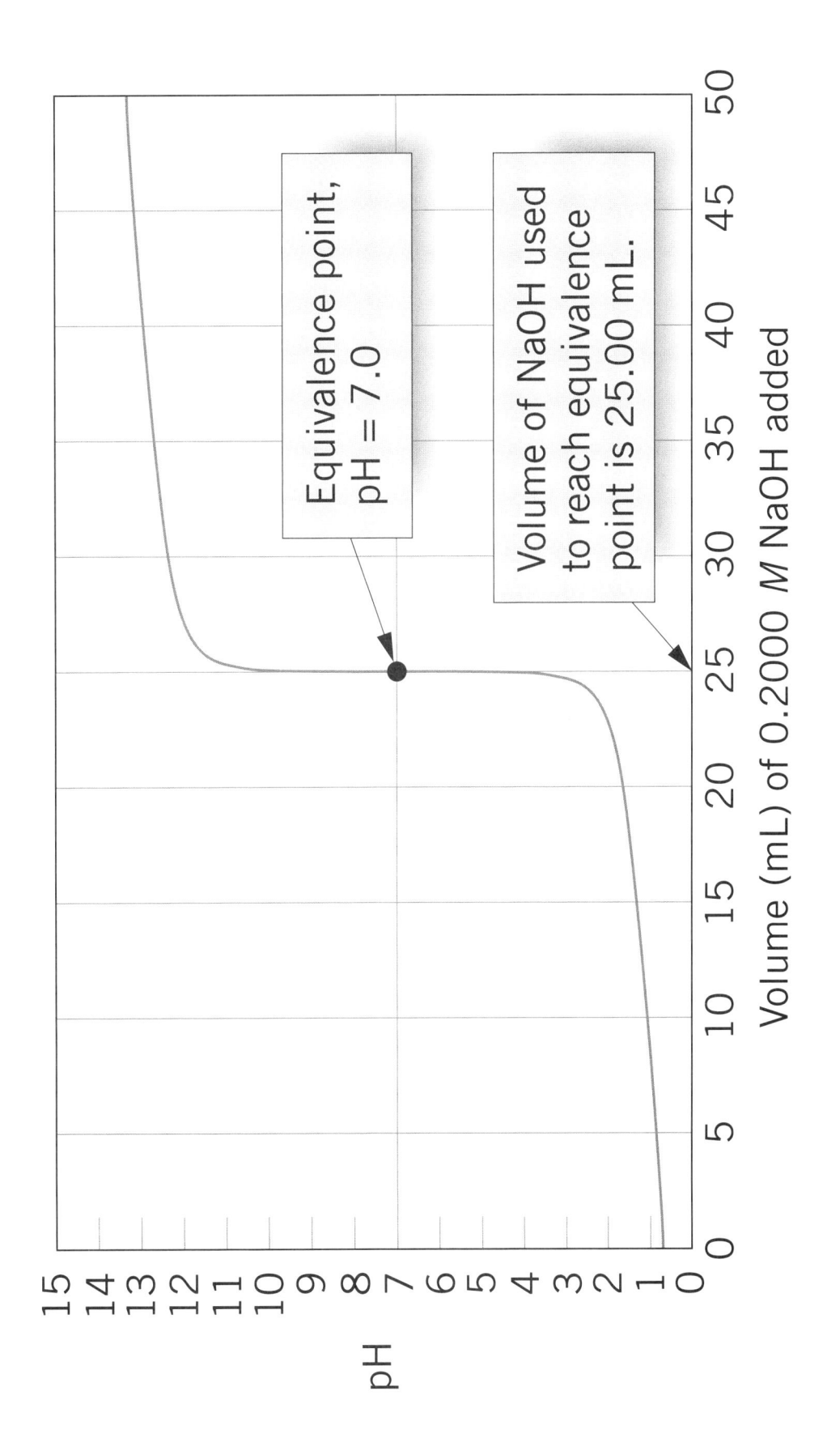

FIGURE 16.6

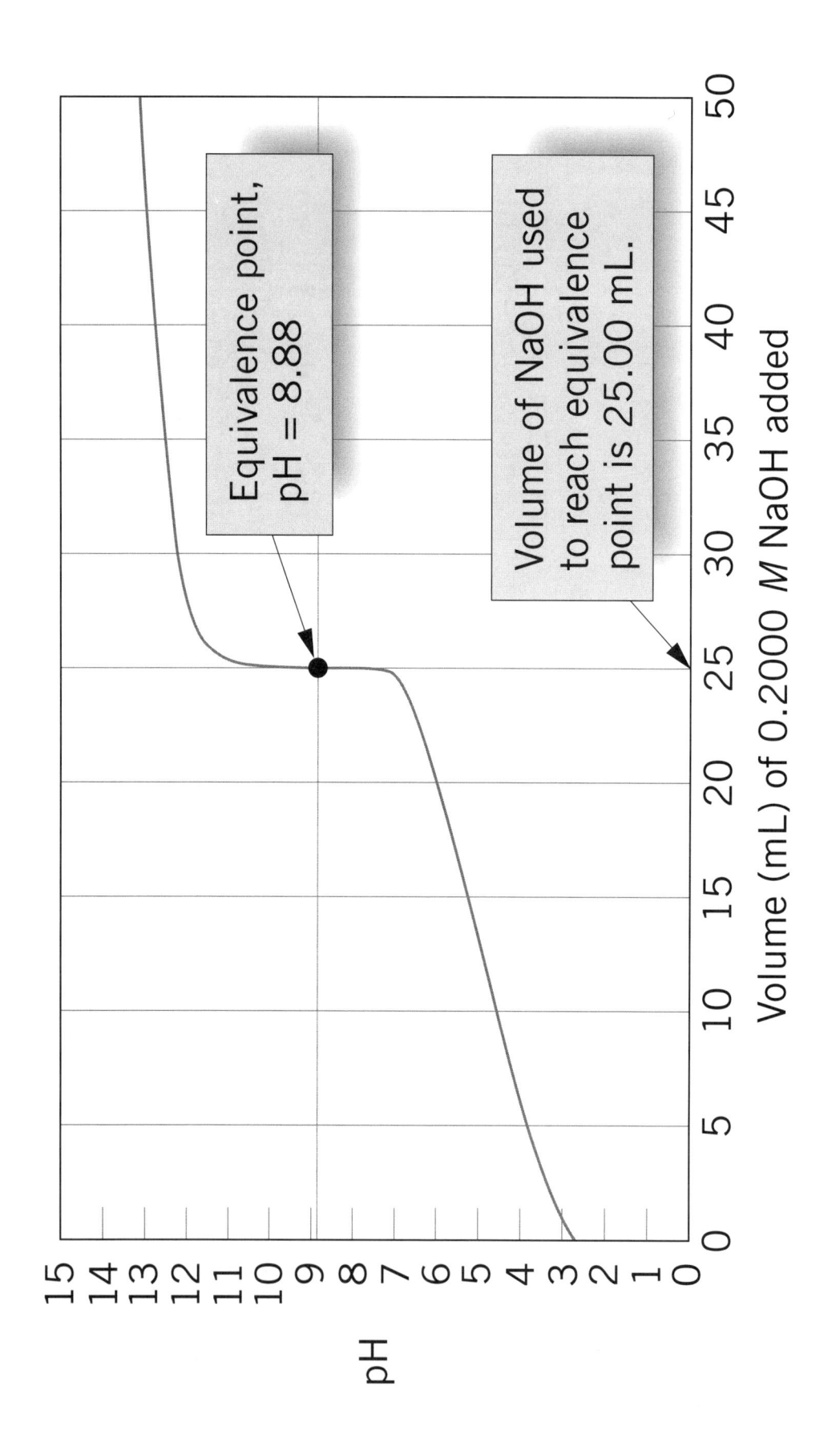

An improbable way for bricks to fall

A more probable way for bricks to fall

(a)

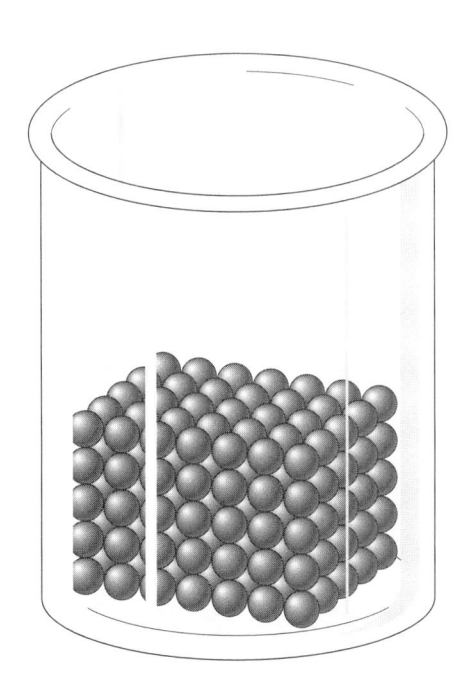

Highly ordered solid – low probability

This change tends to be spontaneous.

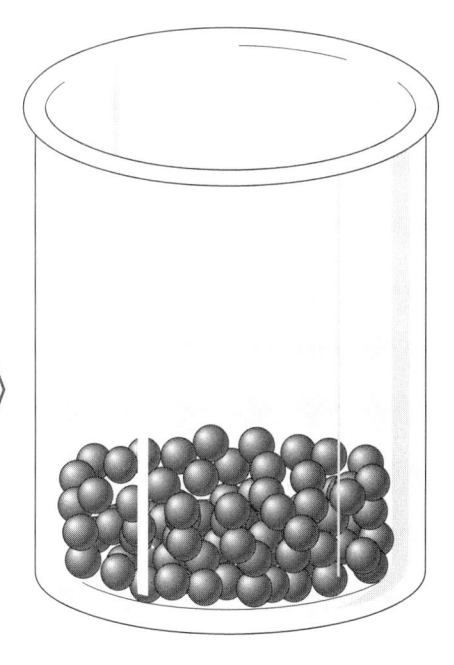

Disordered liquid – higher probability

(b)

FIGURE 18.3

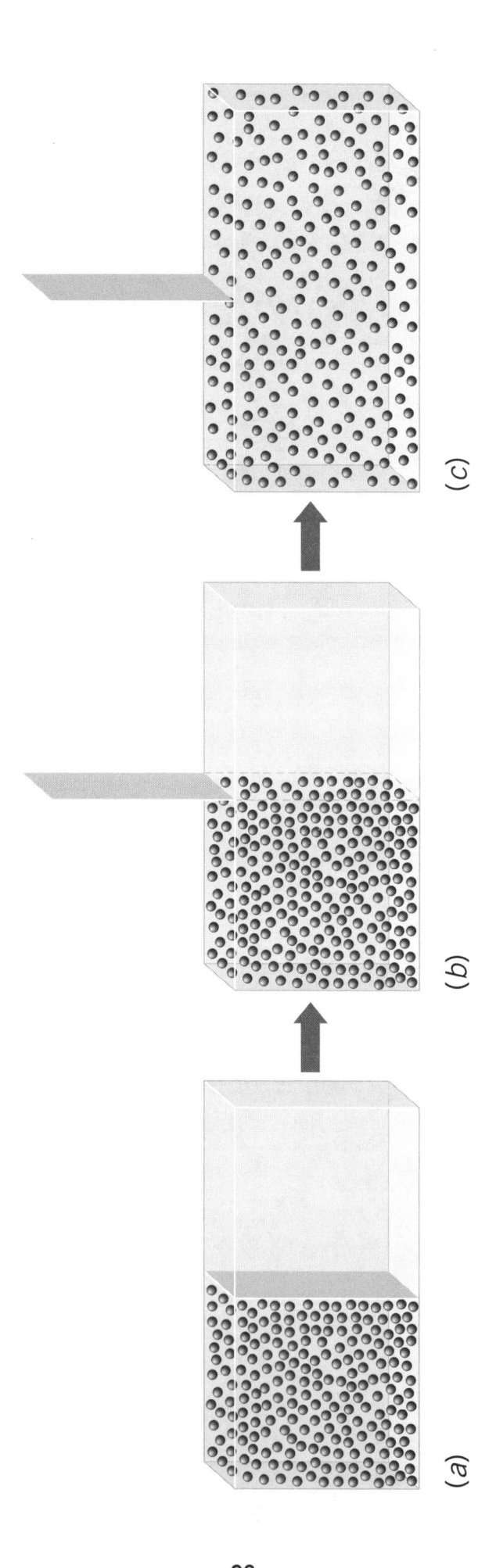

(a)

(b)

(c)

FIGURE 18.4

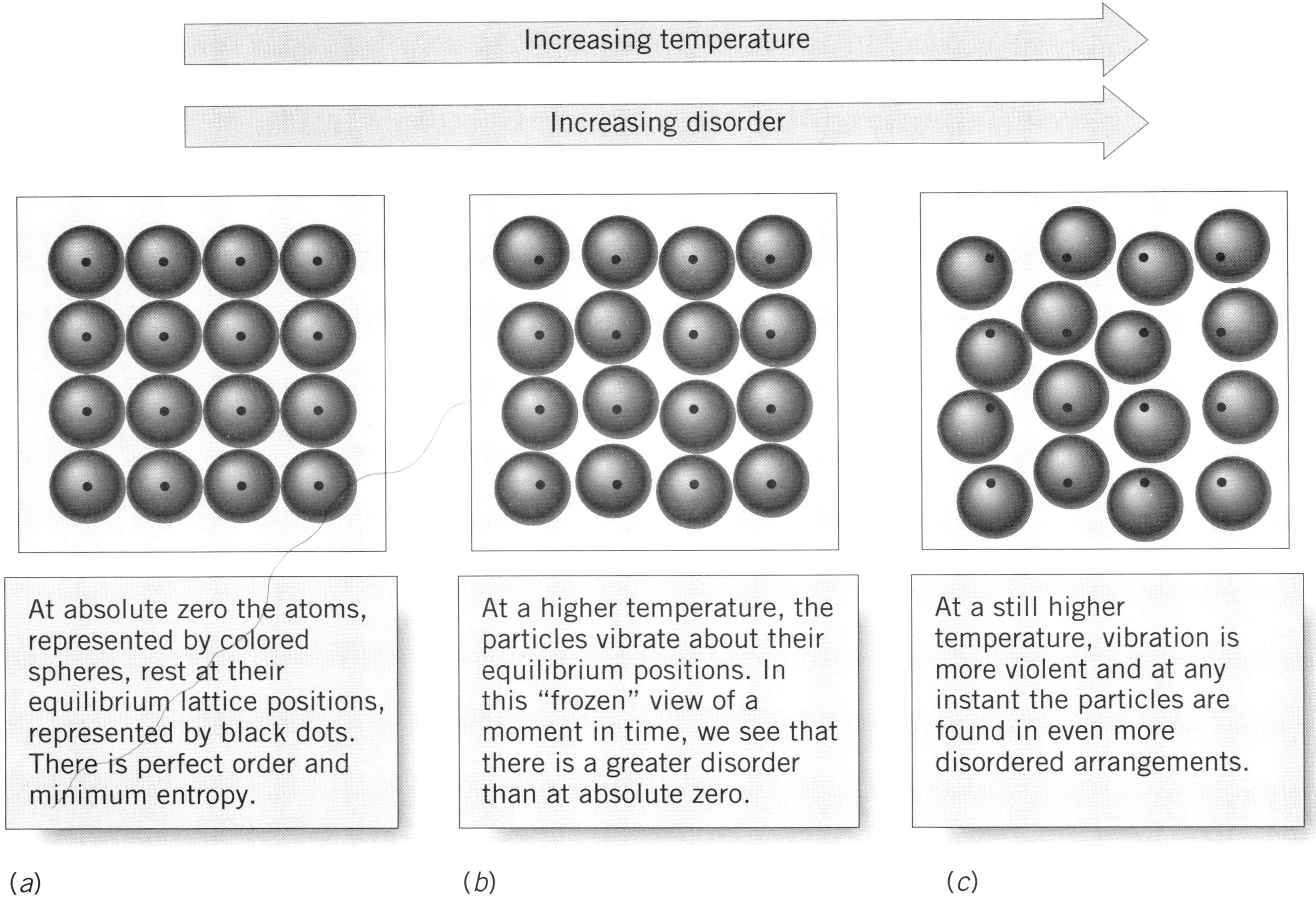

Increasing temperature

Increasing disorder

(a) At absolute zero the atoms, represented by colored spheres, rest at their equilibrium lattice positions, represented by black dots. There is perfect order and minimum entropy.

(b) At a higher temperature, the particles vibrate about their equilibrium positions. In this "frozen" view of a moment in time, we see that there is a greater disorder than at absolute zero.

(c) At a still higher temperature, vibration is more violent and at any instant the particles are found in even more disordered arrangements.

FIGURE 18.5

FIGURE 18.6

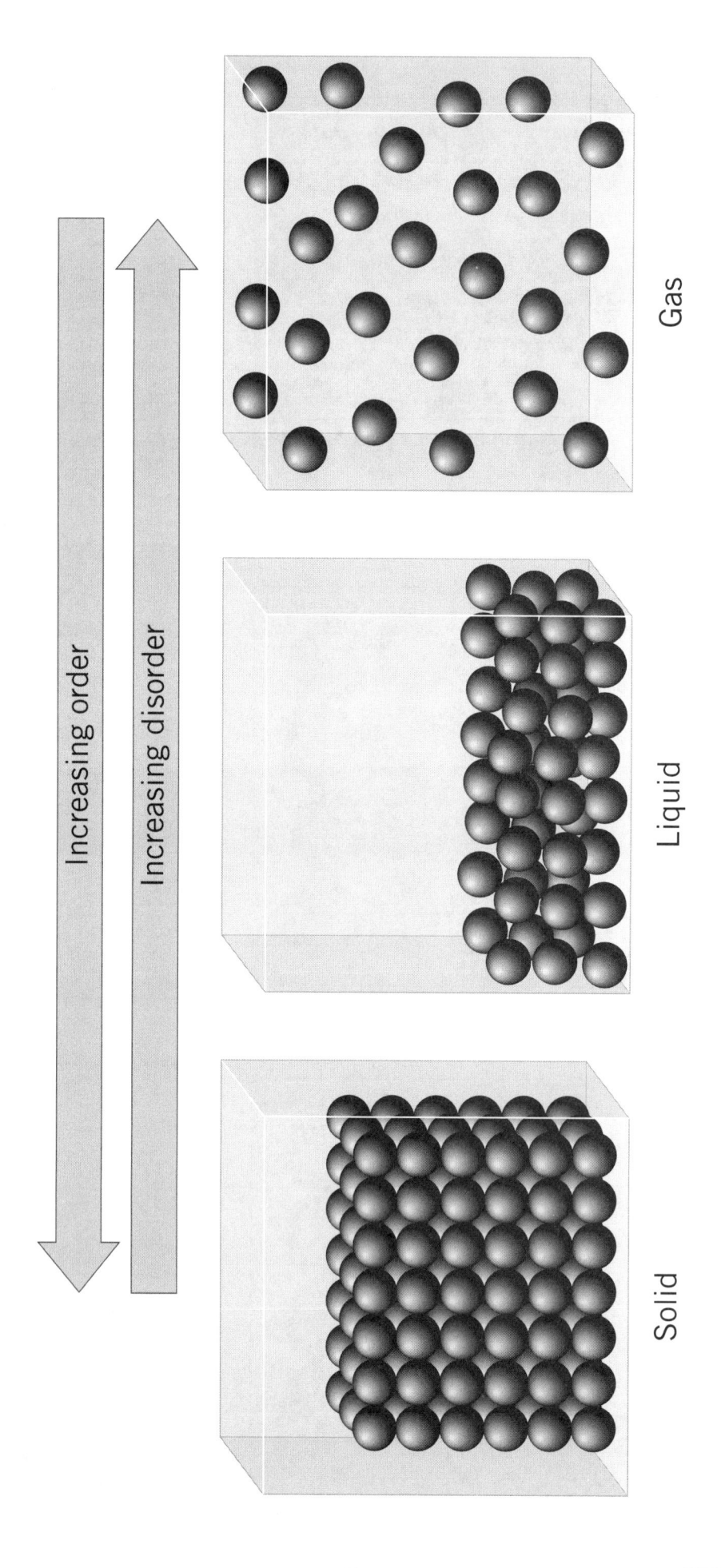

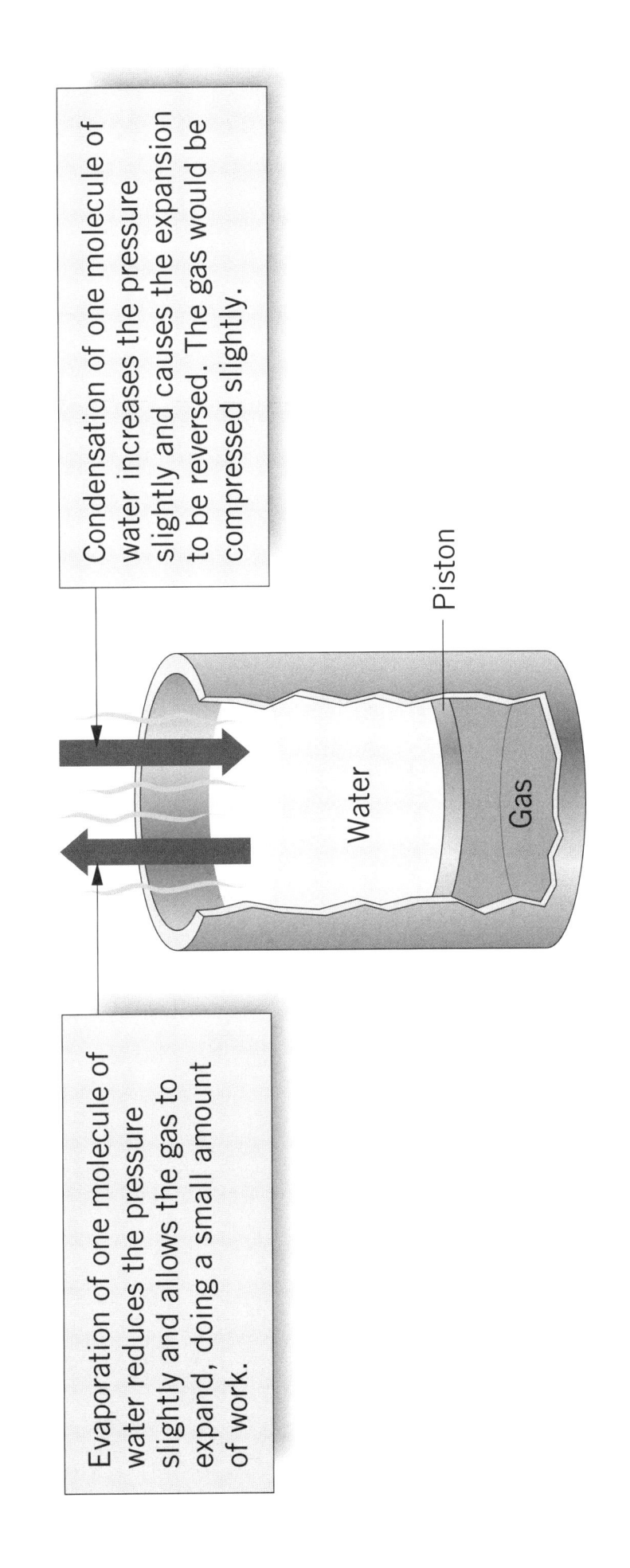

Condensation of one molecule of water increases the pressure slightly and causes the expansion to be reversed. The gas would be compressed slightly.

Evaporation of one molecule of water reduces the pressure slightly and allows the gas to expand, doing a small amount of work.

Piston

Water

Gas

FIGURE 18.10

Cl₂ molecules rise to the surface of the liquid.

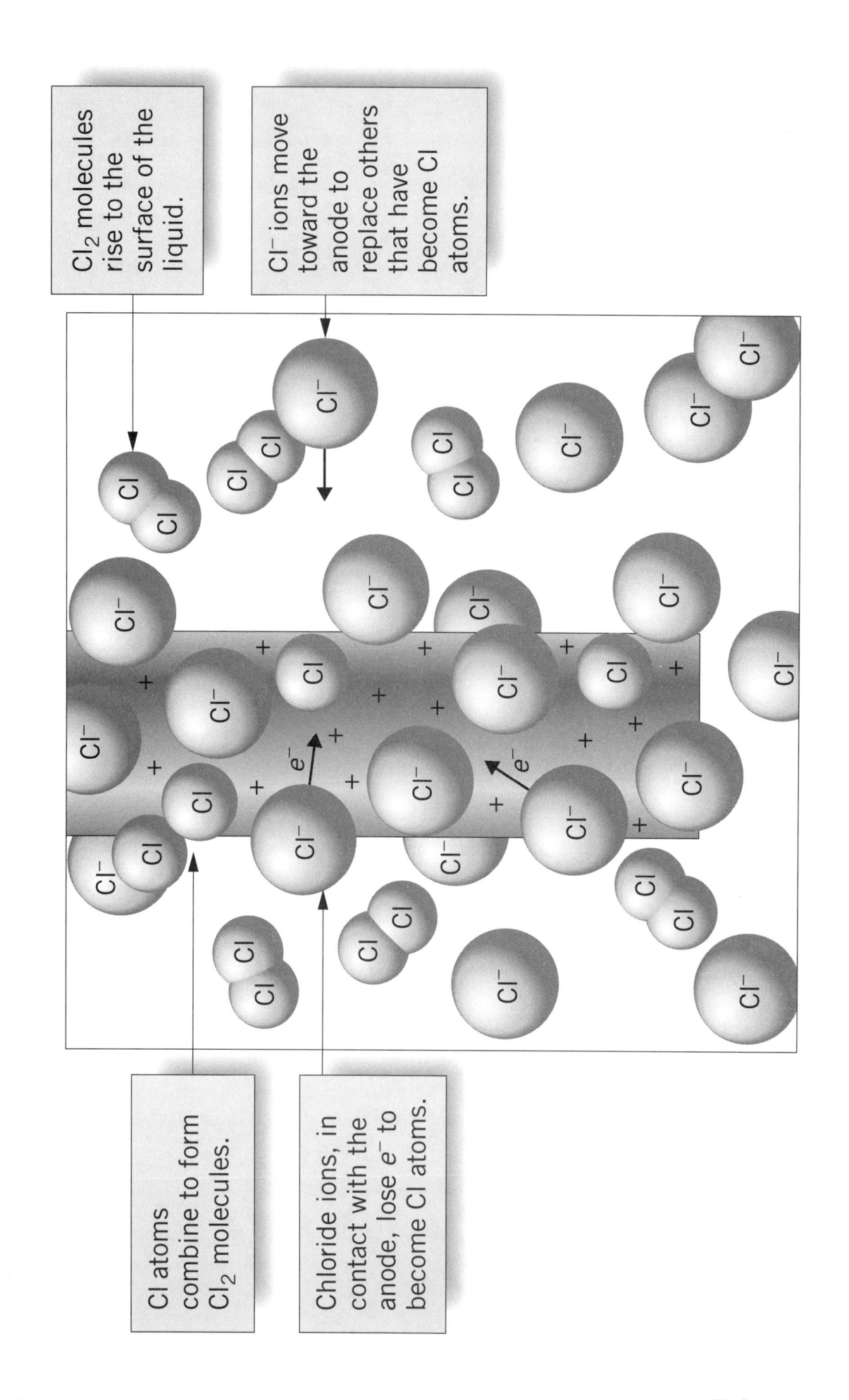

Cl⁻ ions move toward the anode to replace others that have become Cl atoms.

Cl atoms combine to form Cl₂ molecules.

Chloride ions, in contact with the anode, lose e^- to become Cl atoms.

FIGURE 19.2

FIGURE 19.7

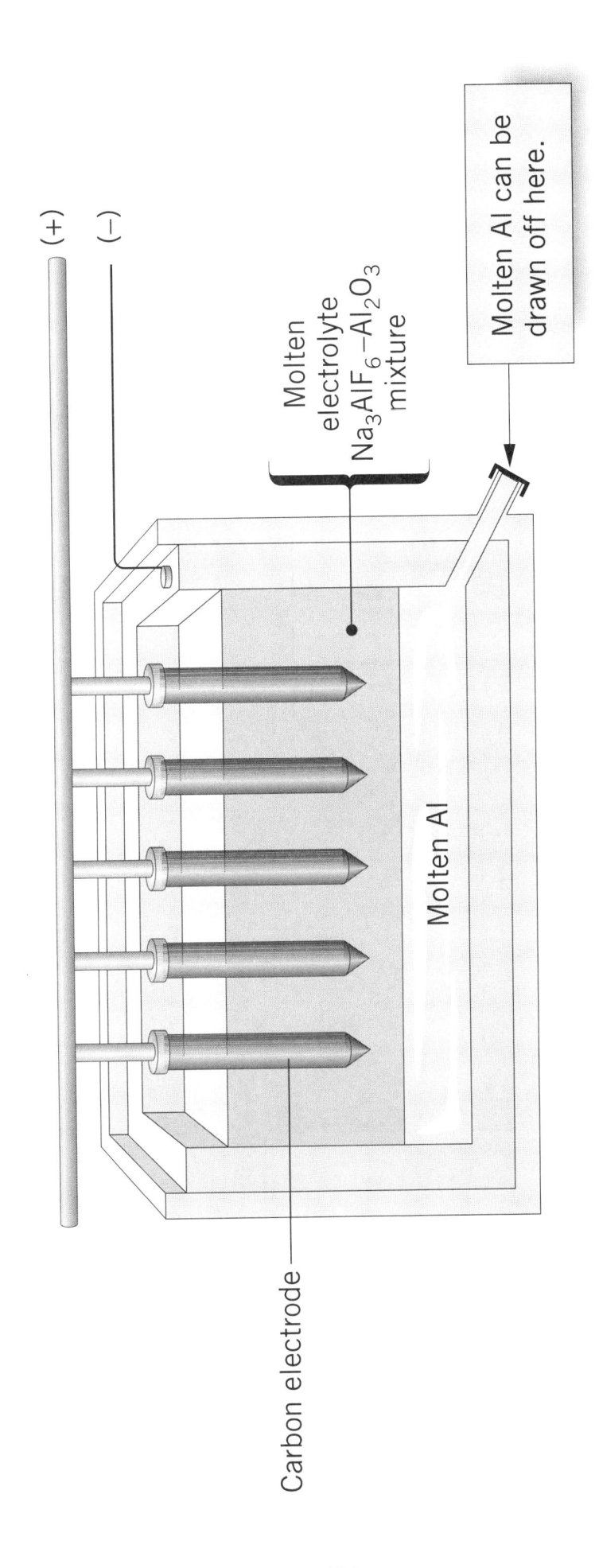

FIGURE 19.8

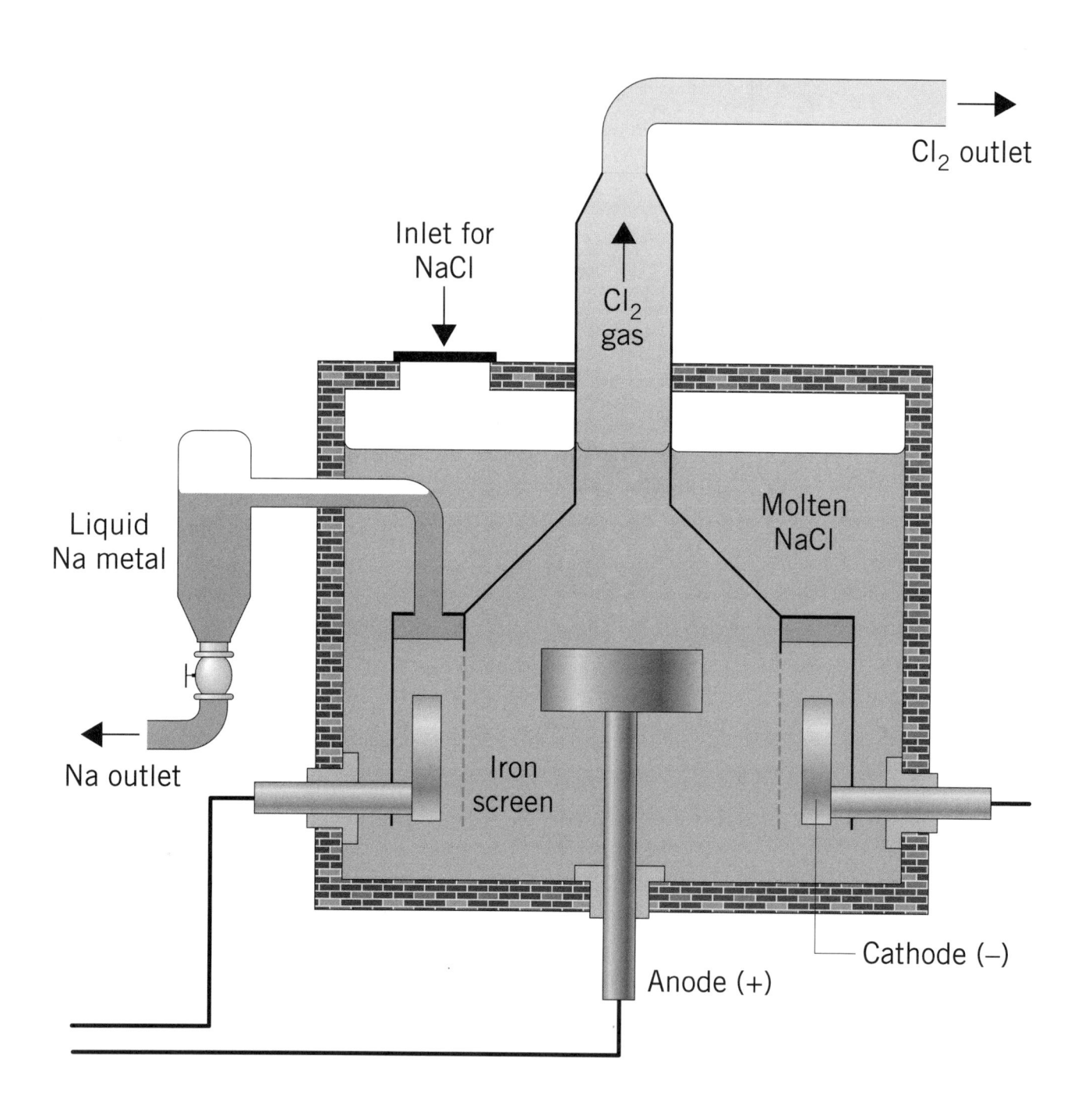

FIGURE 19.11

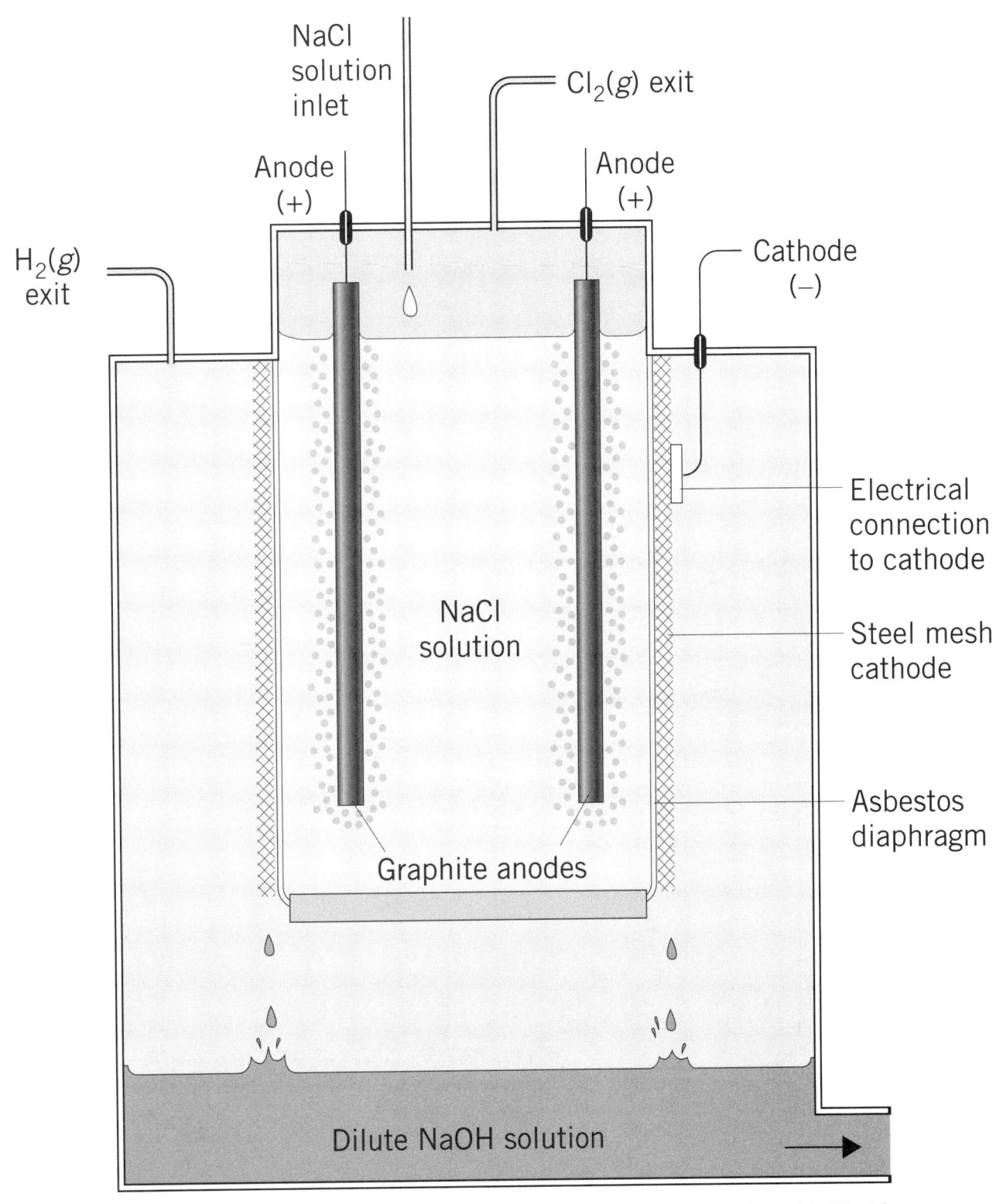

FIGURE 19.15

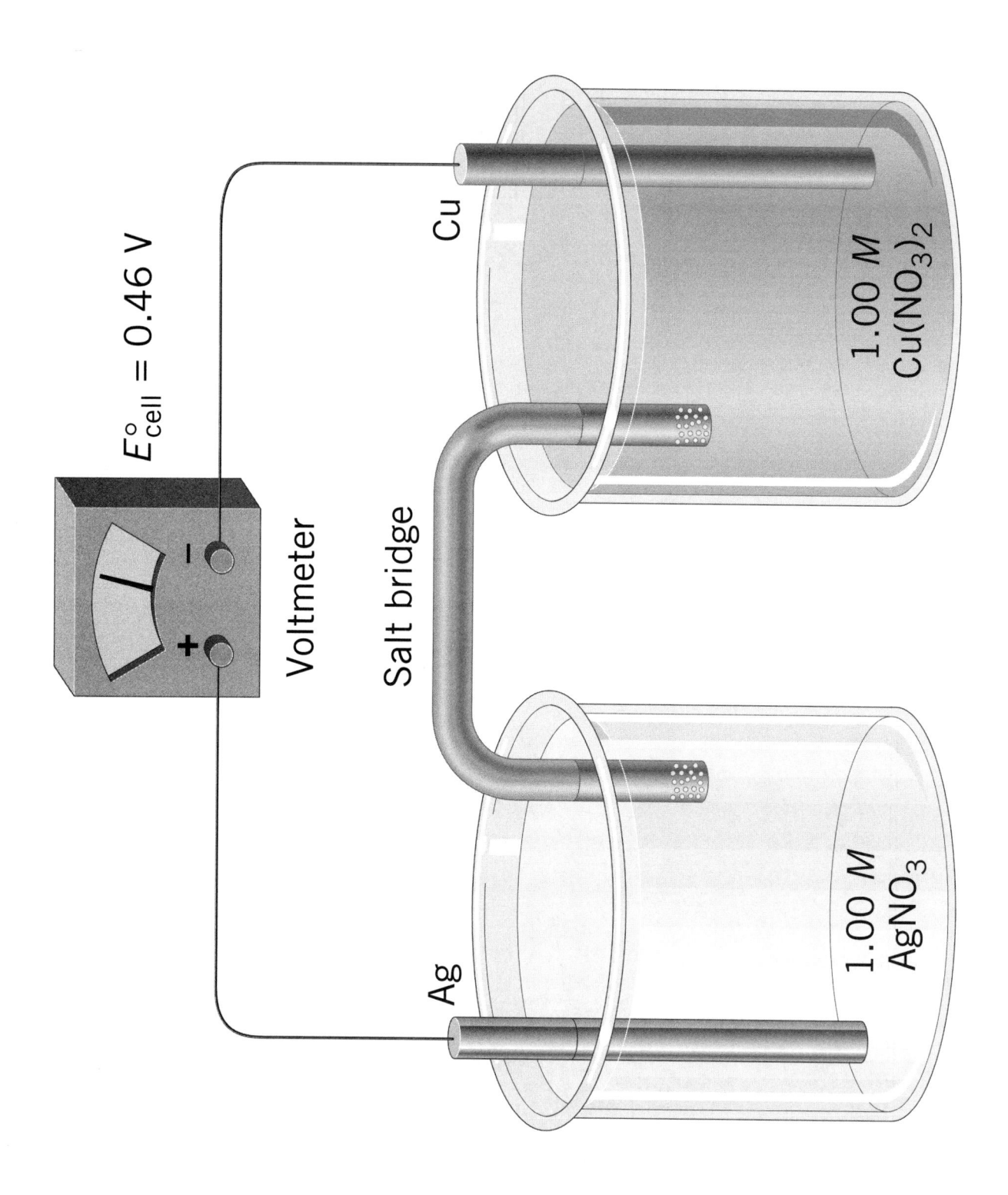

$E°_{H^+}$ = 0.00 V

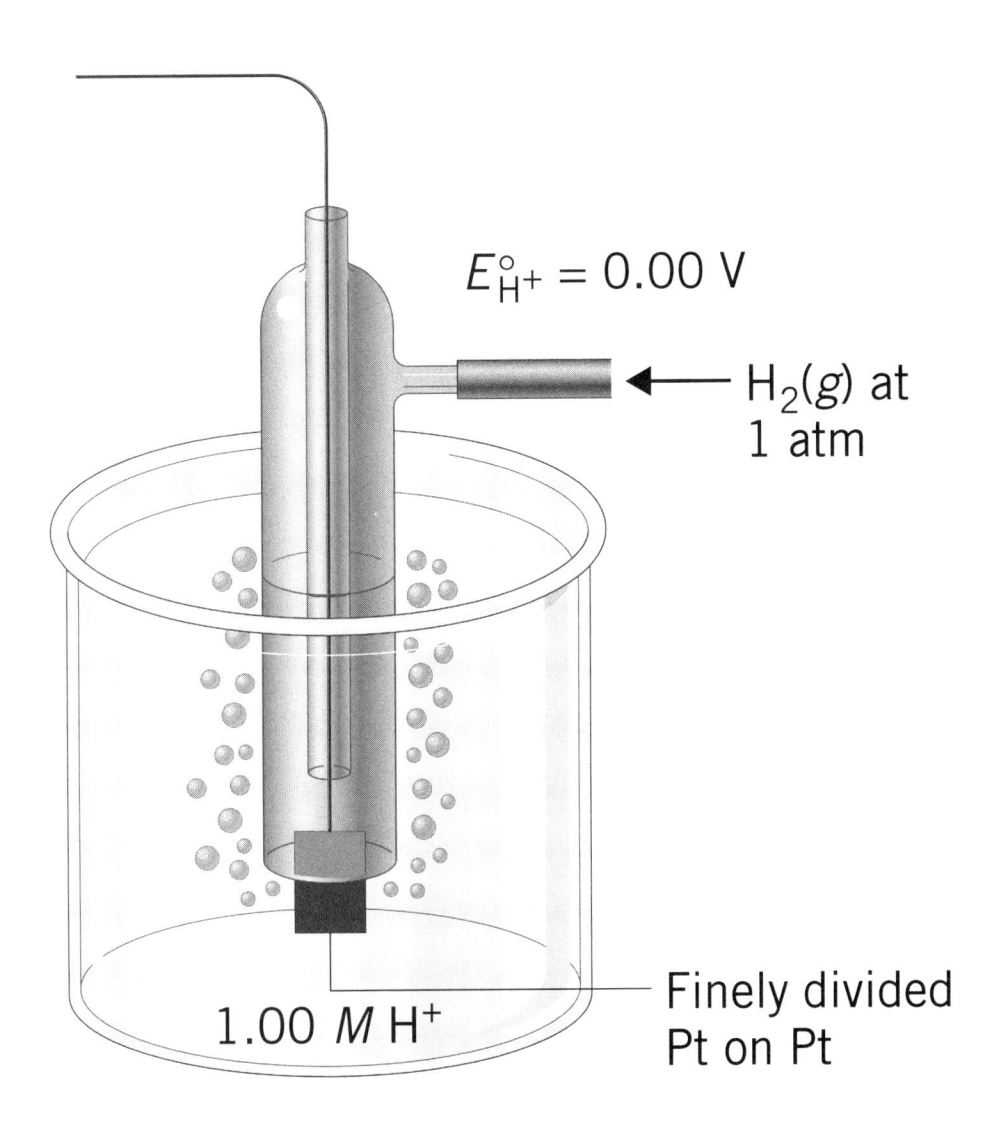

← H$_2$(g) at 1 atm

1.00 M H$^+$

Finely divided Pt on Pt

FIGURE 19.16

Isomer II

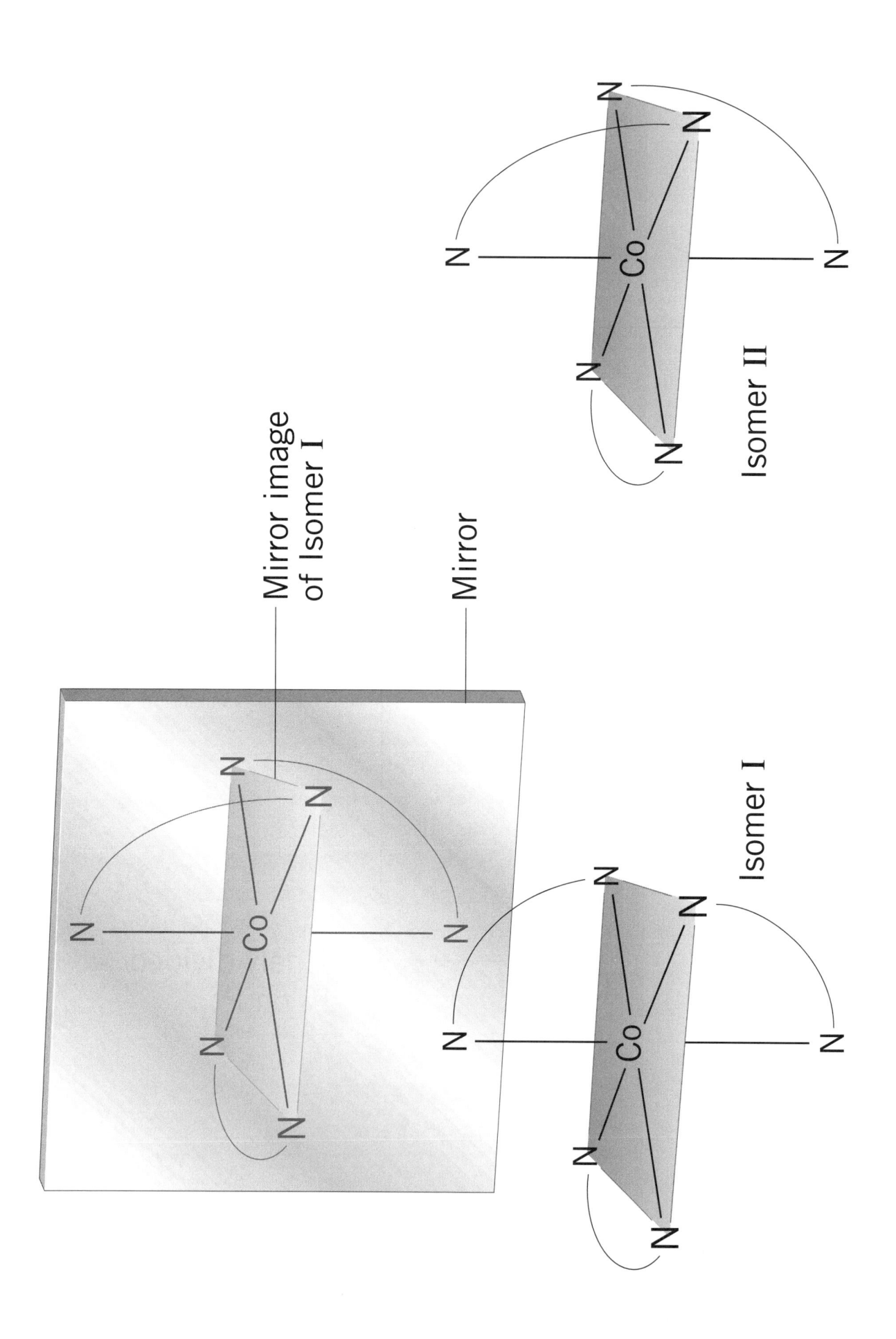

Mirror image of Isomer I

Mirror

Isomer I

FIGURE 20.10

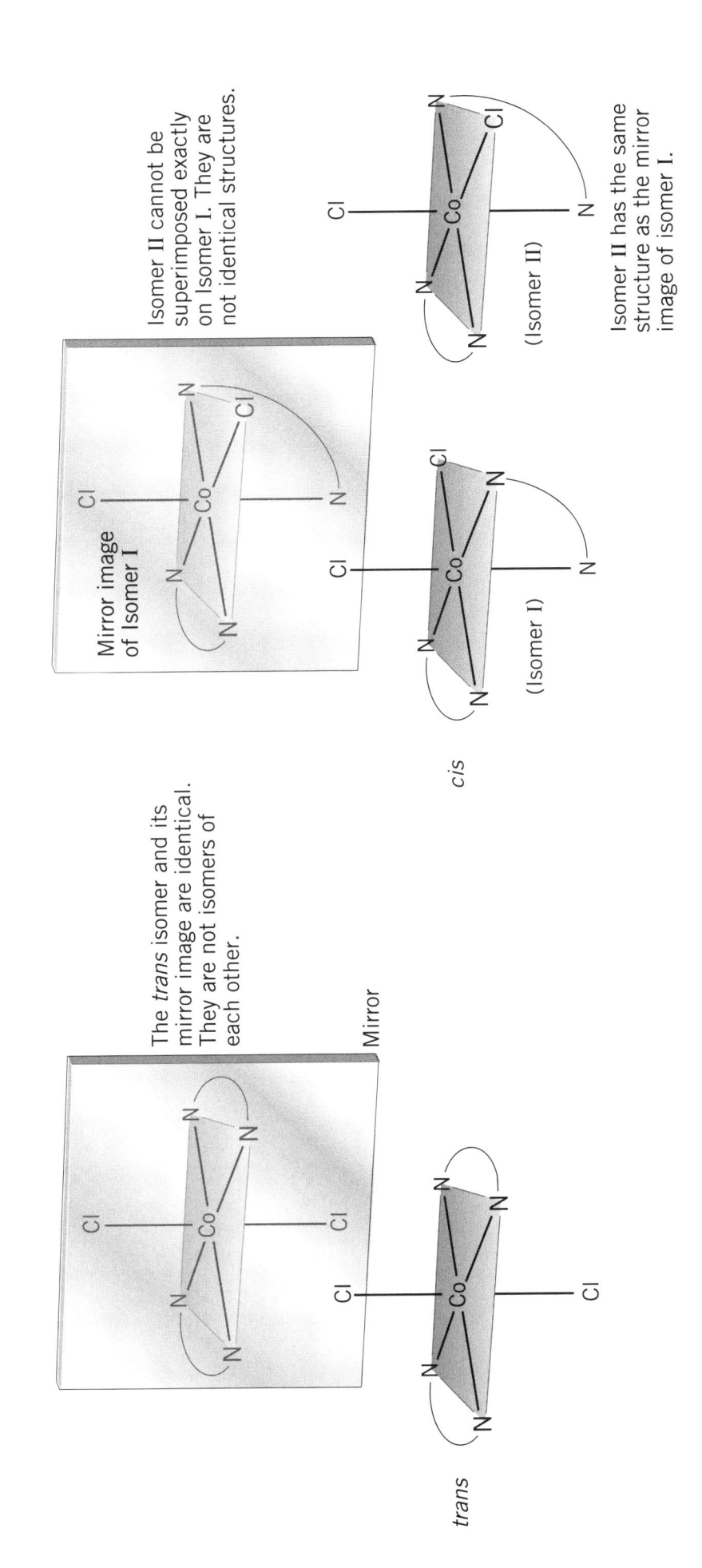

Isomer II cannot be superimposed exactly on Isomer I. They are not identical structures.

(Isomer II)

Isomer II has the same structure as the mirror image of isomer I.

Mirror image of Isomer I

(Isomer I)

cis

The *trans* isomer and its mirror image are identical. They are not isomers of each other.

Mirror

trans

FIGURE 20.11

108

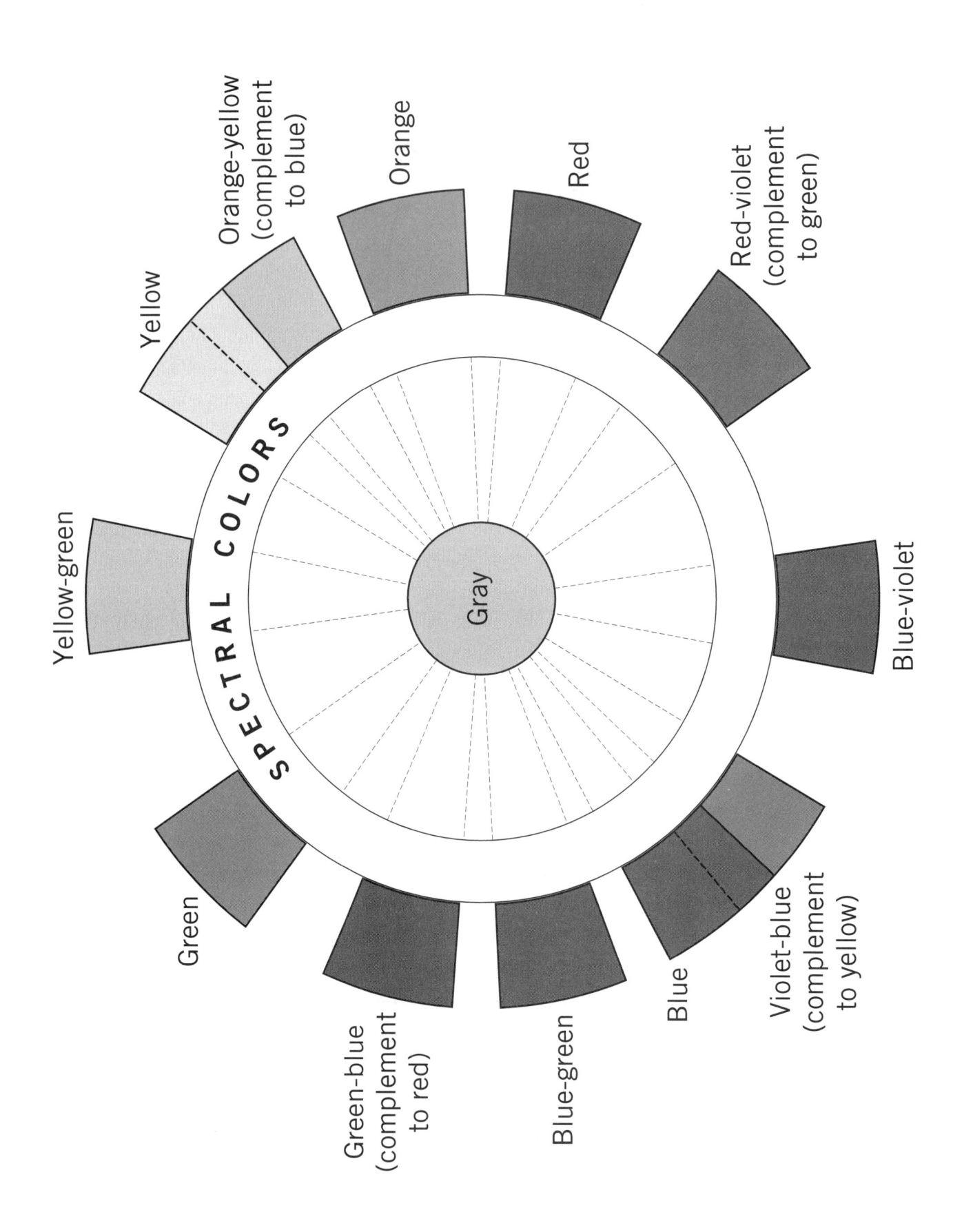

FIGURE 20.18

FIGURE 20.21

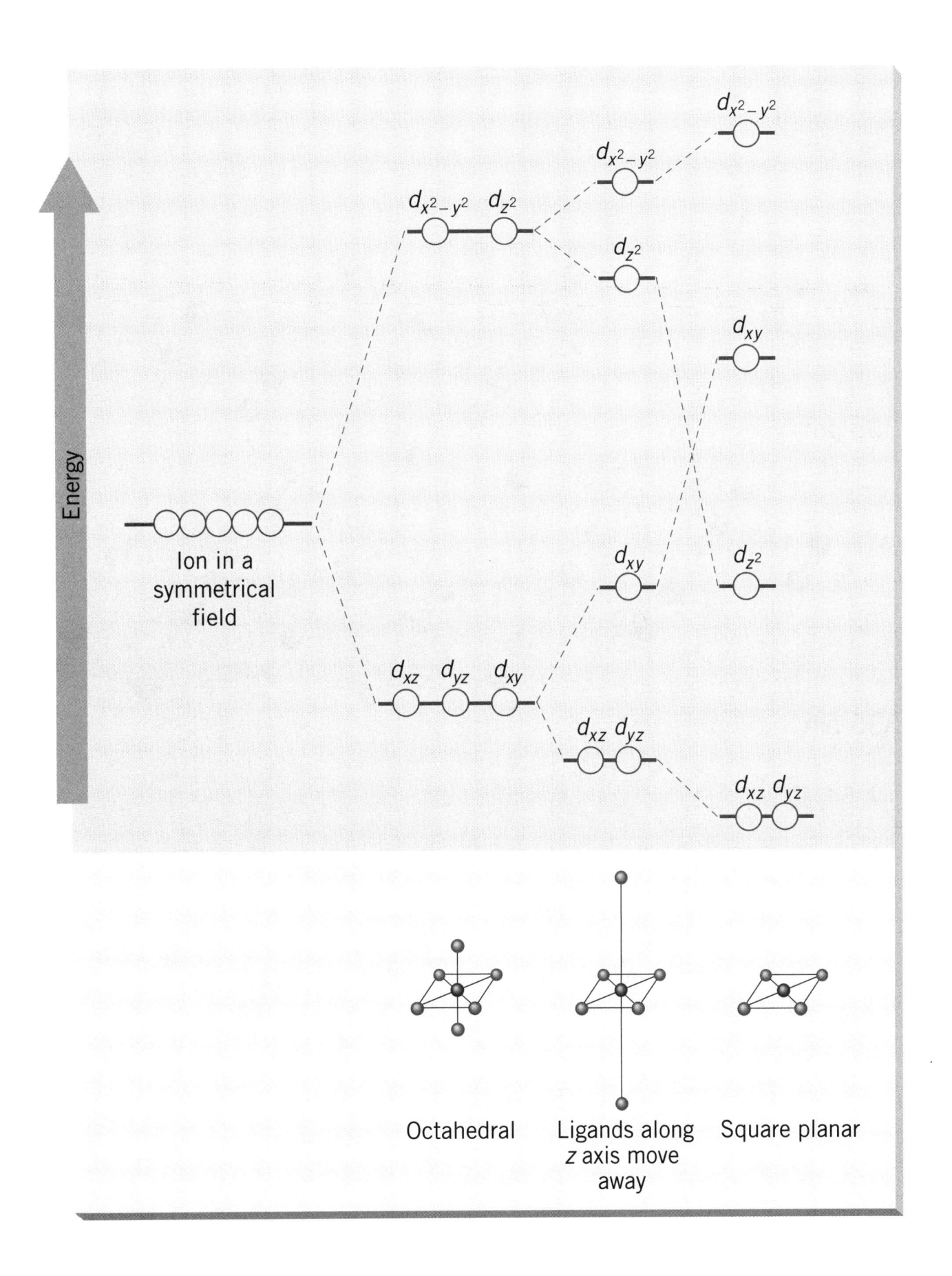

FIGURE 21.6

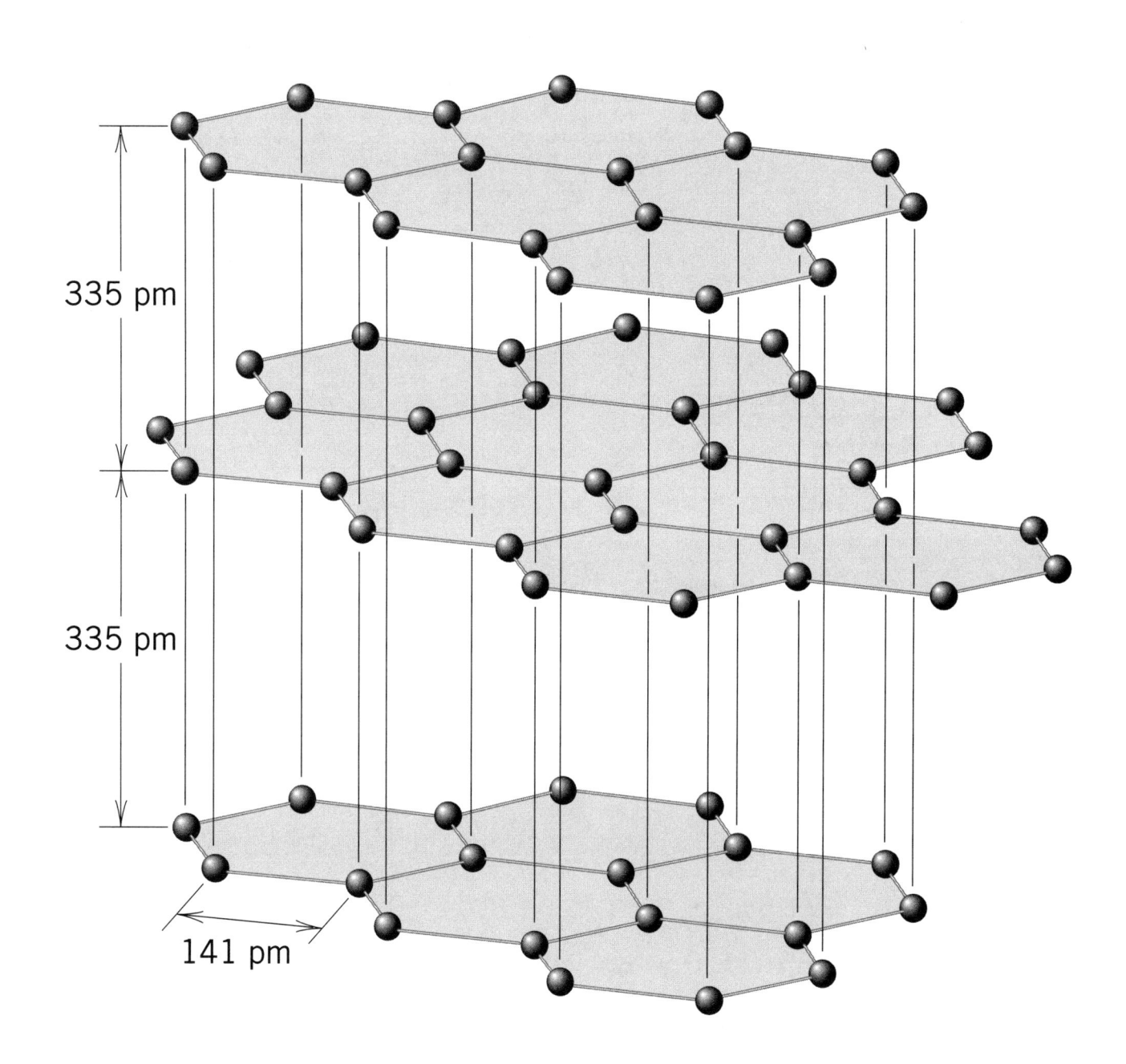

C_{60}

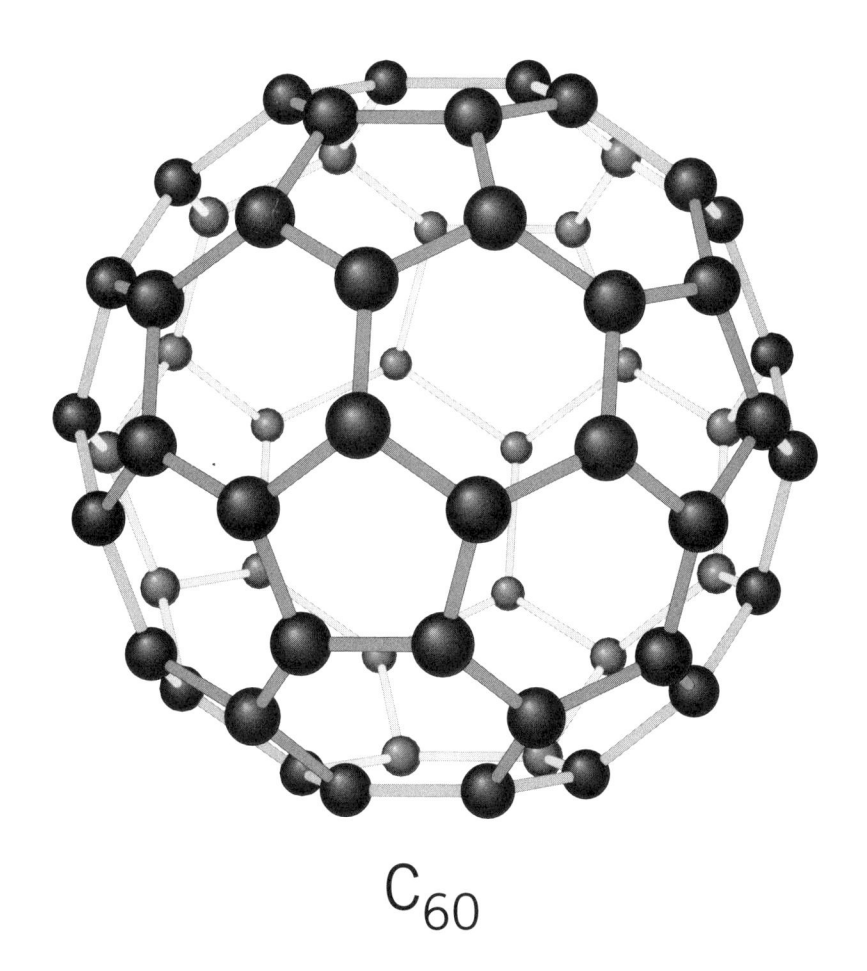

FIGURE 21.7

S8

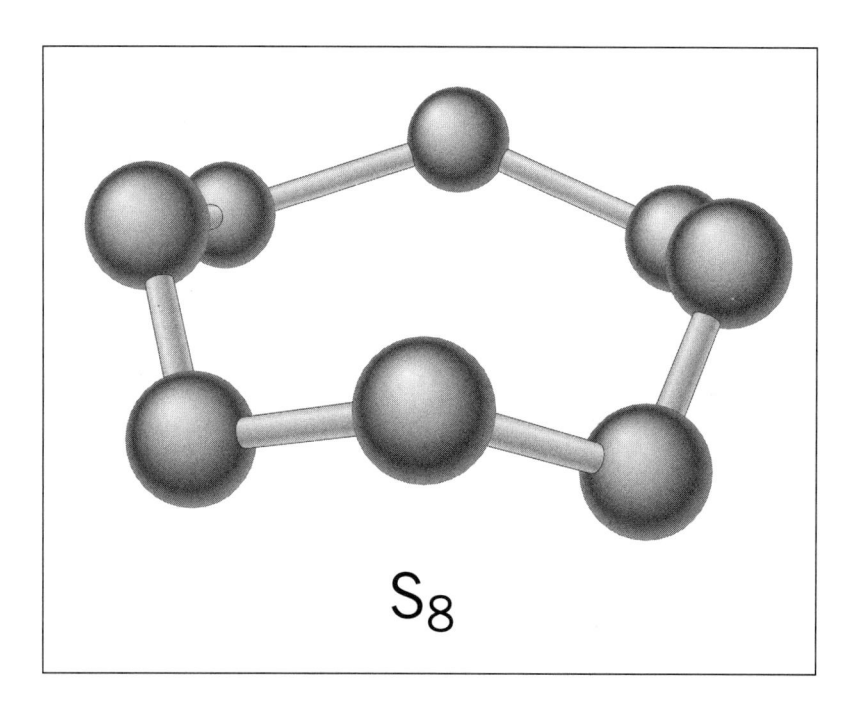

FIGURE 21.9

P_4

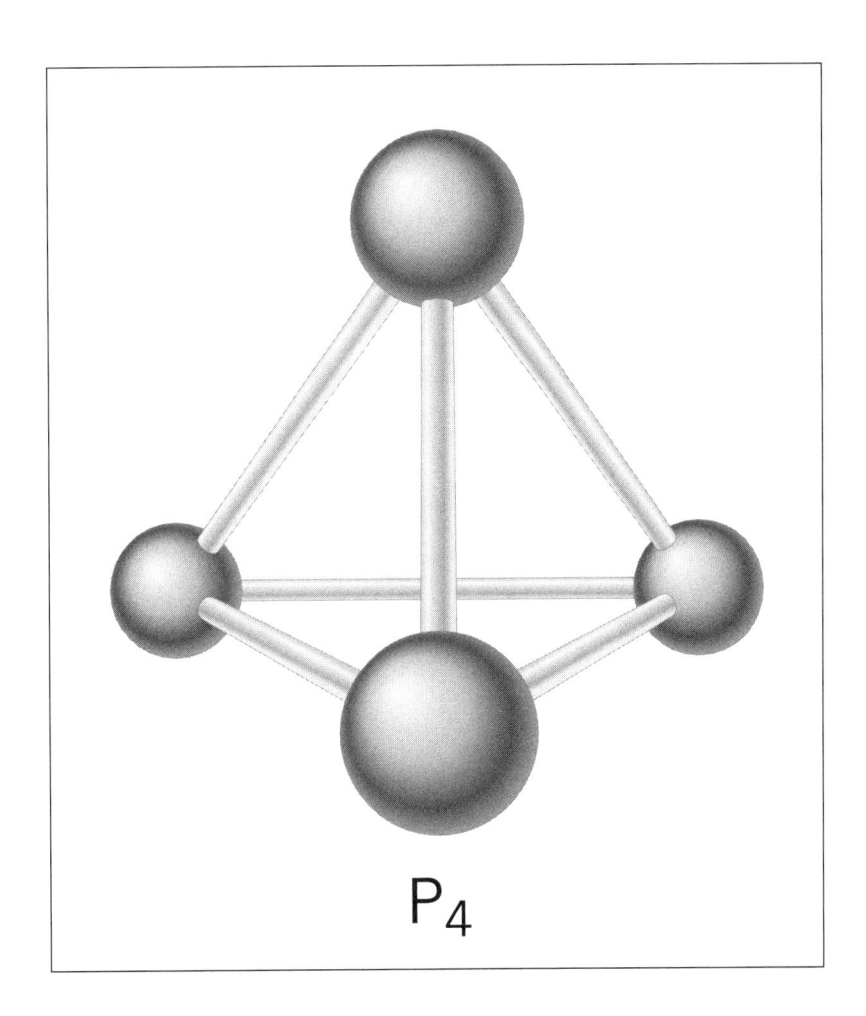

FIGURE 21.10

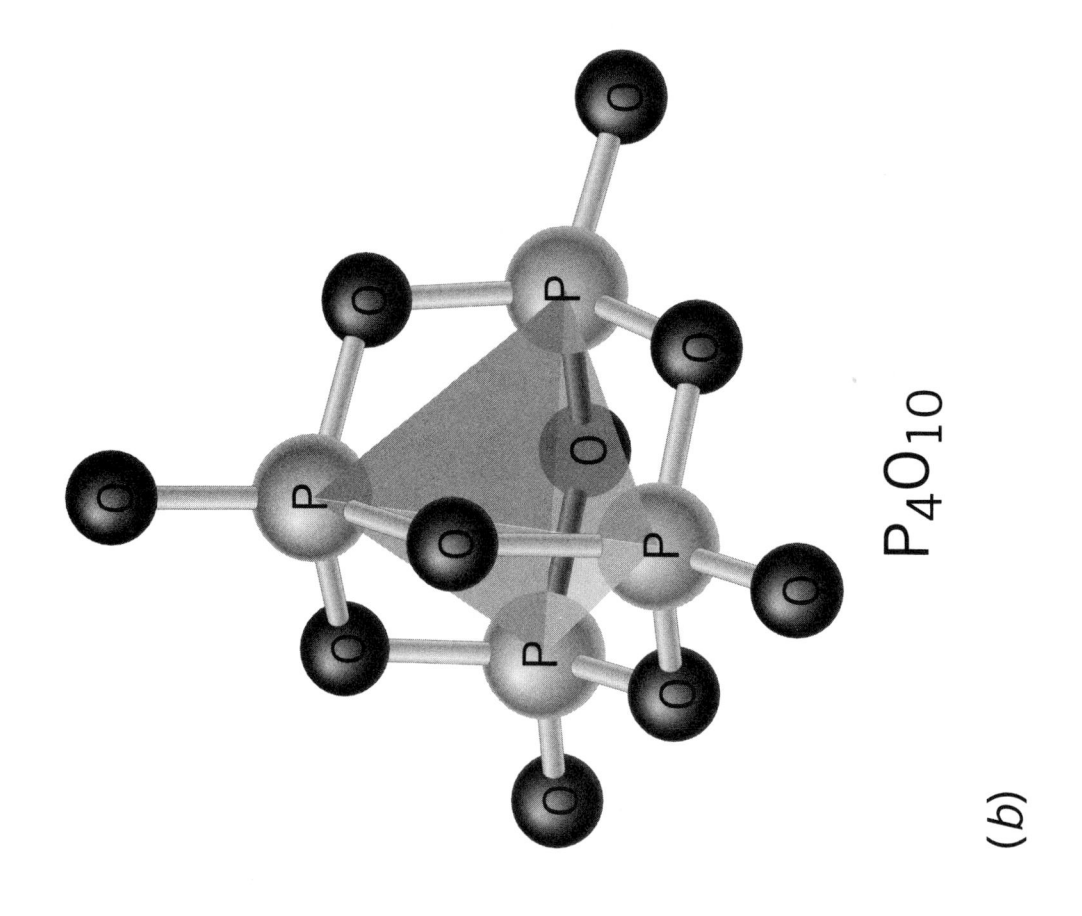

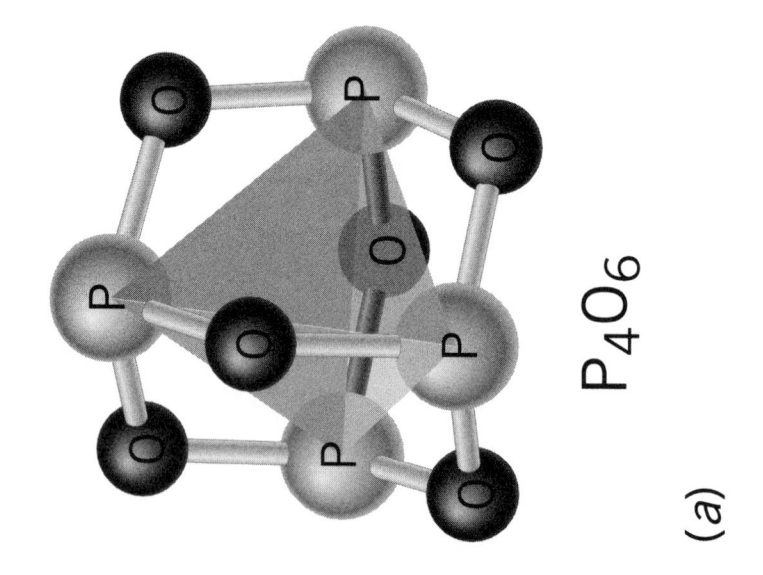

P_4O_{10}

(b)

P_4O_6

(a)

FIGURE 21.16

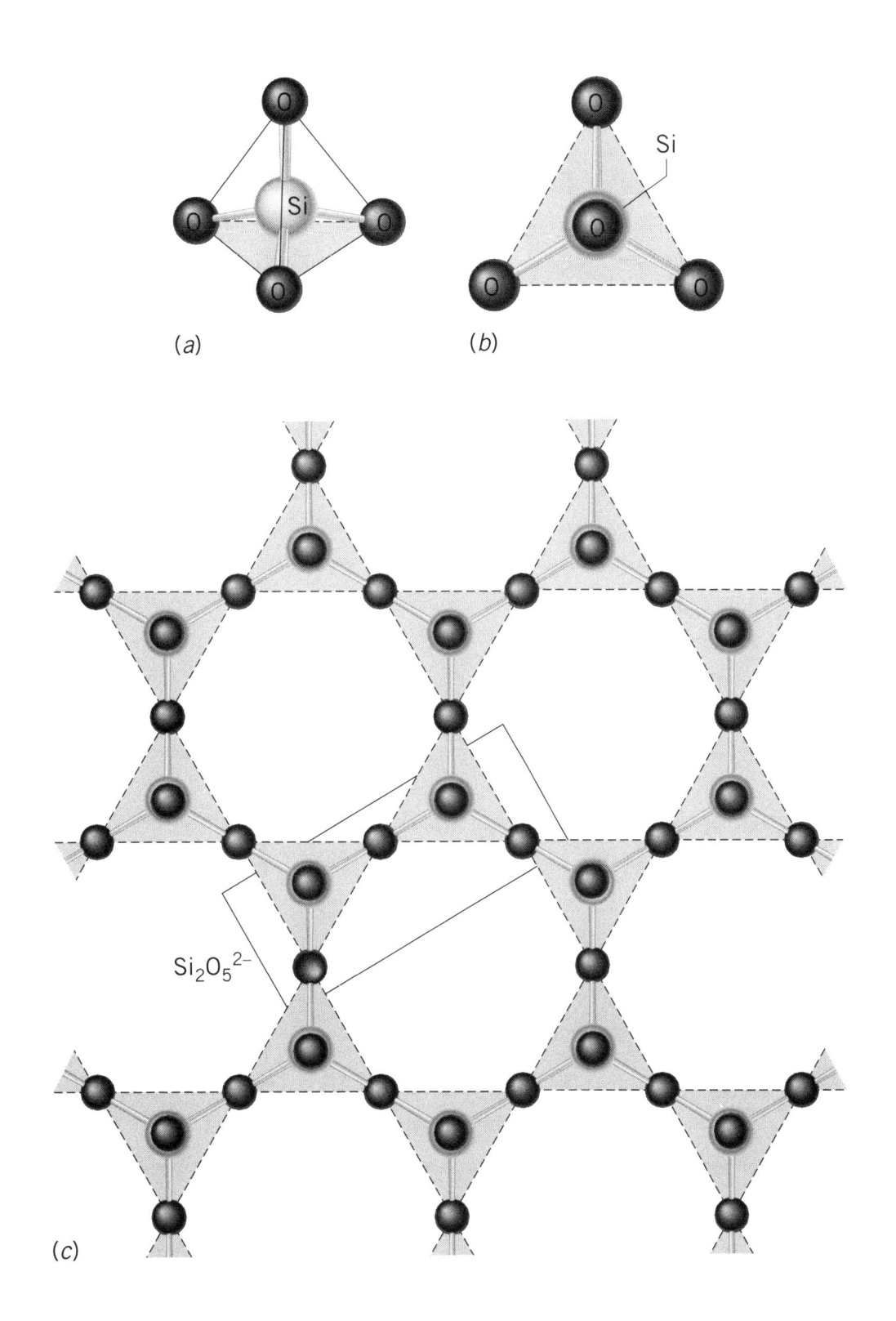

(a)

(b)

Si₂O₅²⁻

(c)

FIGURE 21.22

FIGURE 22.1

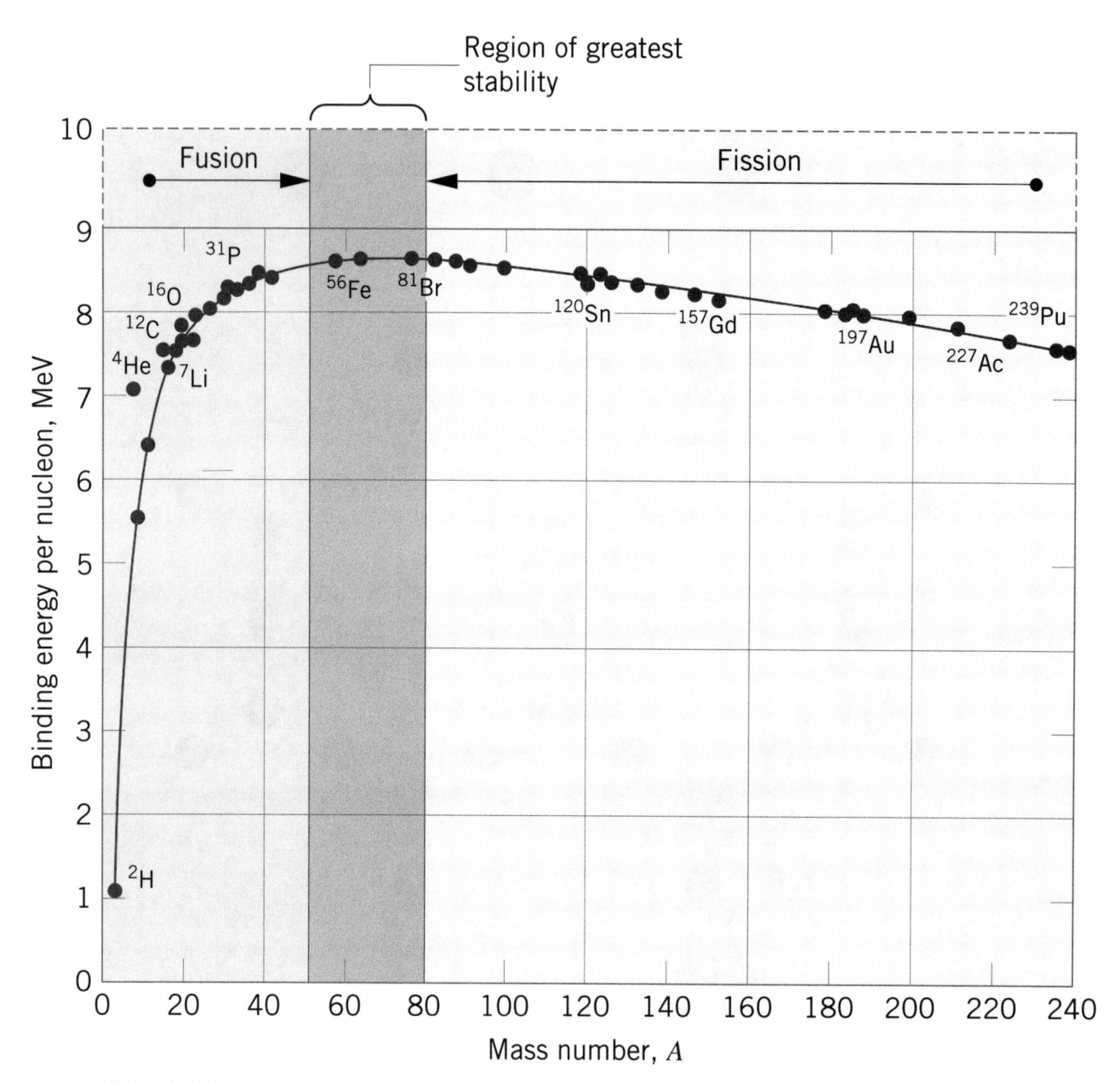

FIGURE 22.8

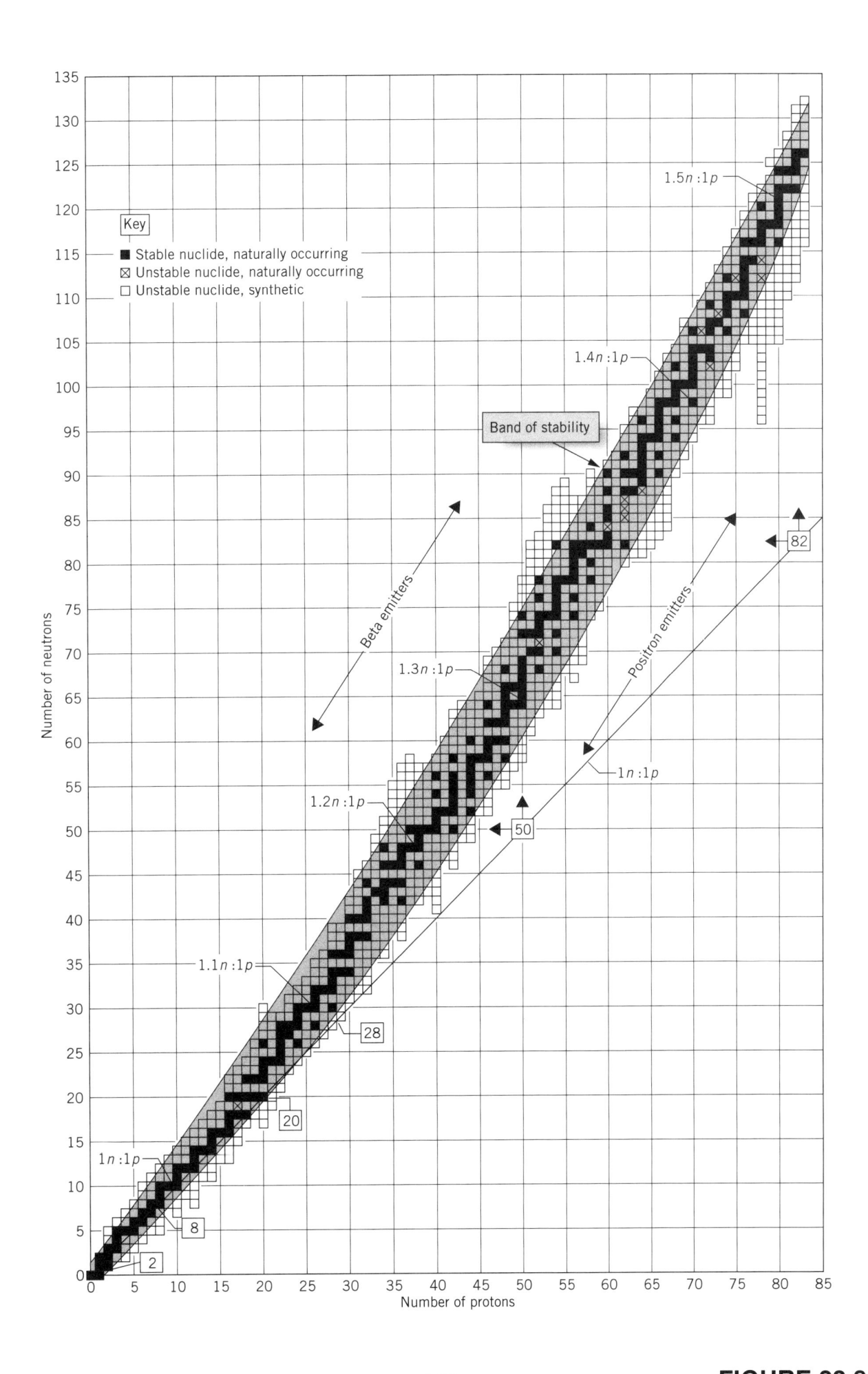

FIGURE 22.9

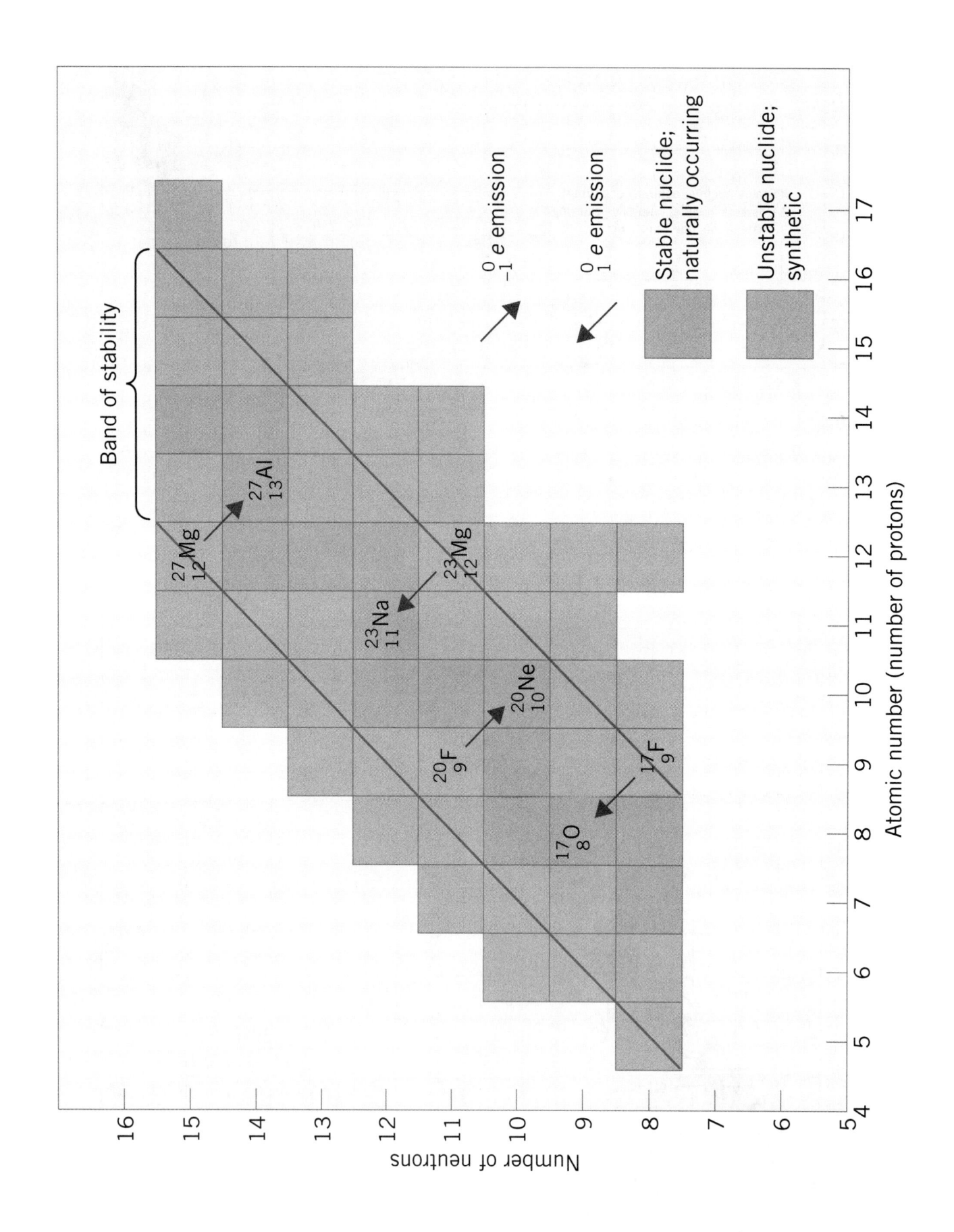

FIGURE 22.16

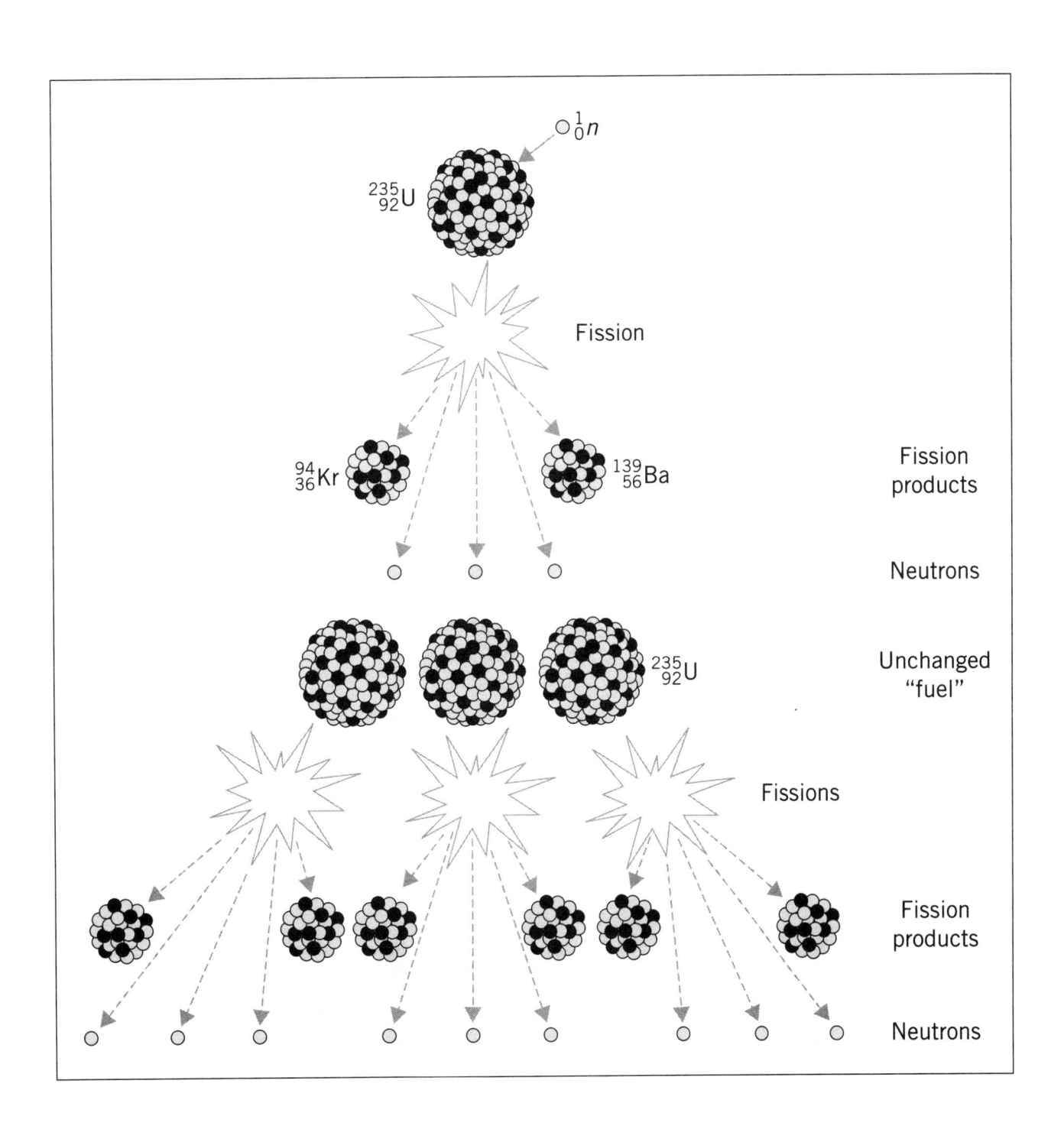

FIGURE 22.17

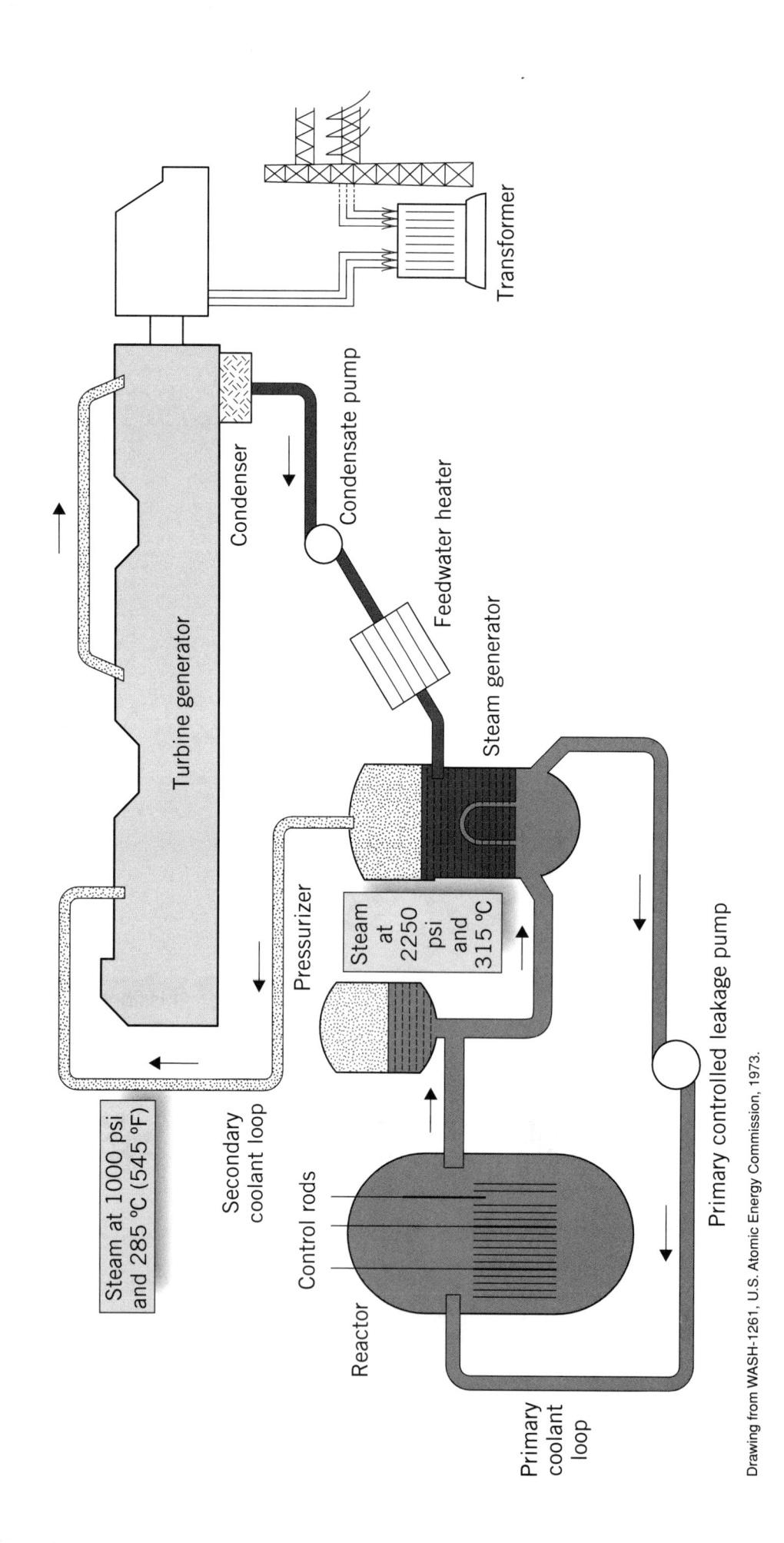

FIGURE 23.7

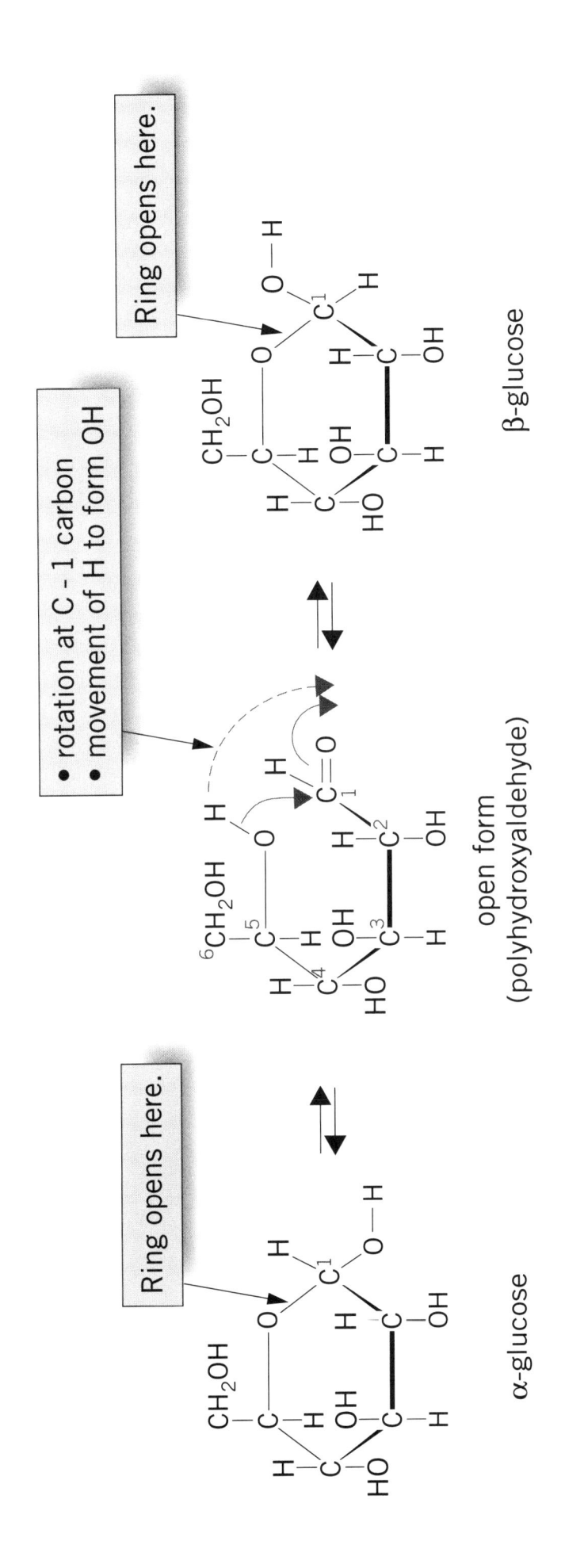

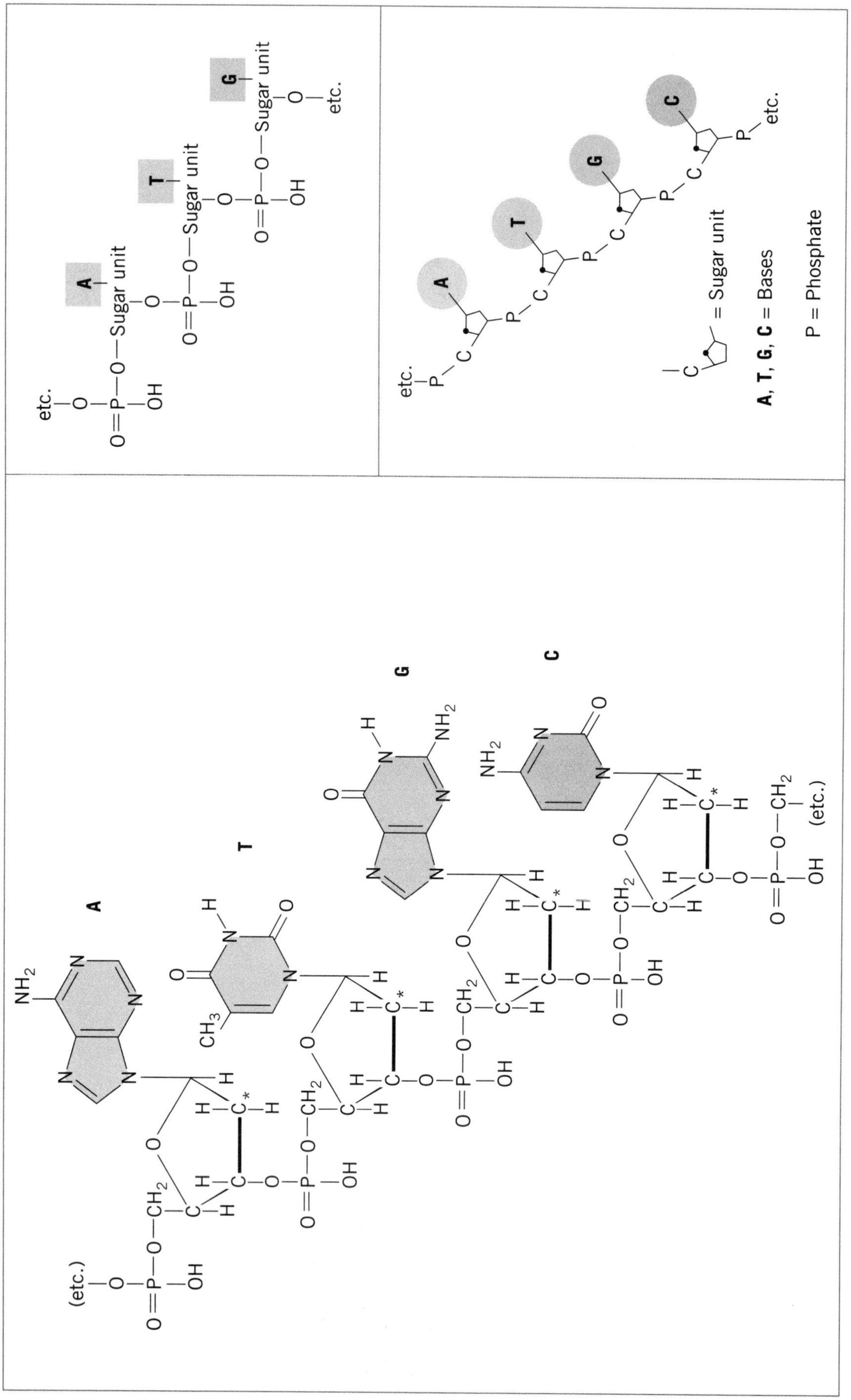

FIGURE 23.11

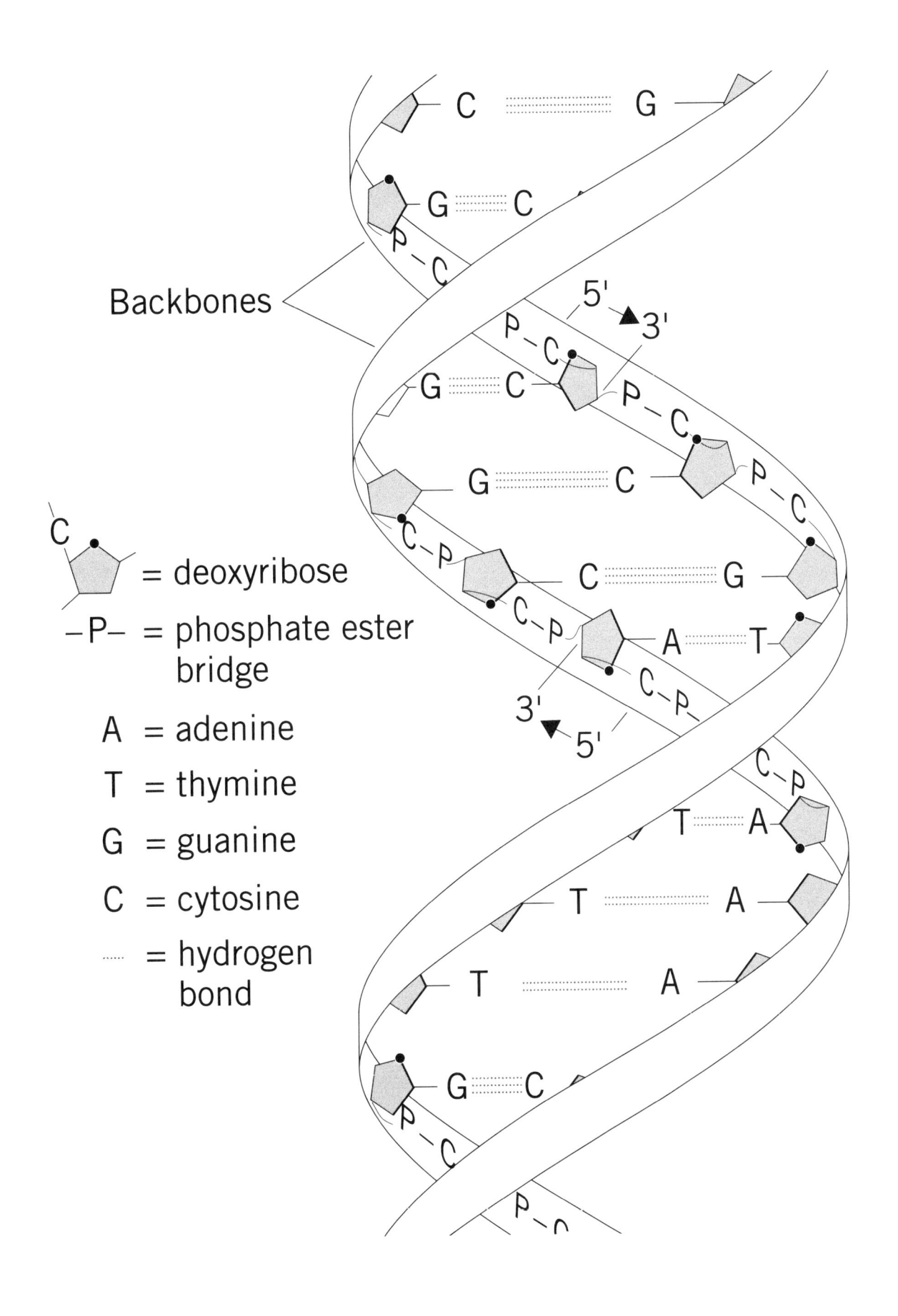

- C ▫ = deoxyribose
- –P– = phosphate ester bridge
- A = adenine
- T = thymine
- G = guanine
- C = cytosine
- ⋯ = hydrogen bond

Backbones

(a)

FIGURE 23.12a

FIGURE 23.13

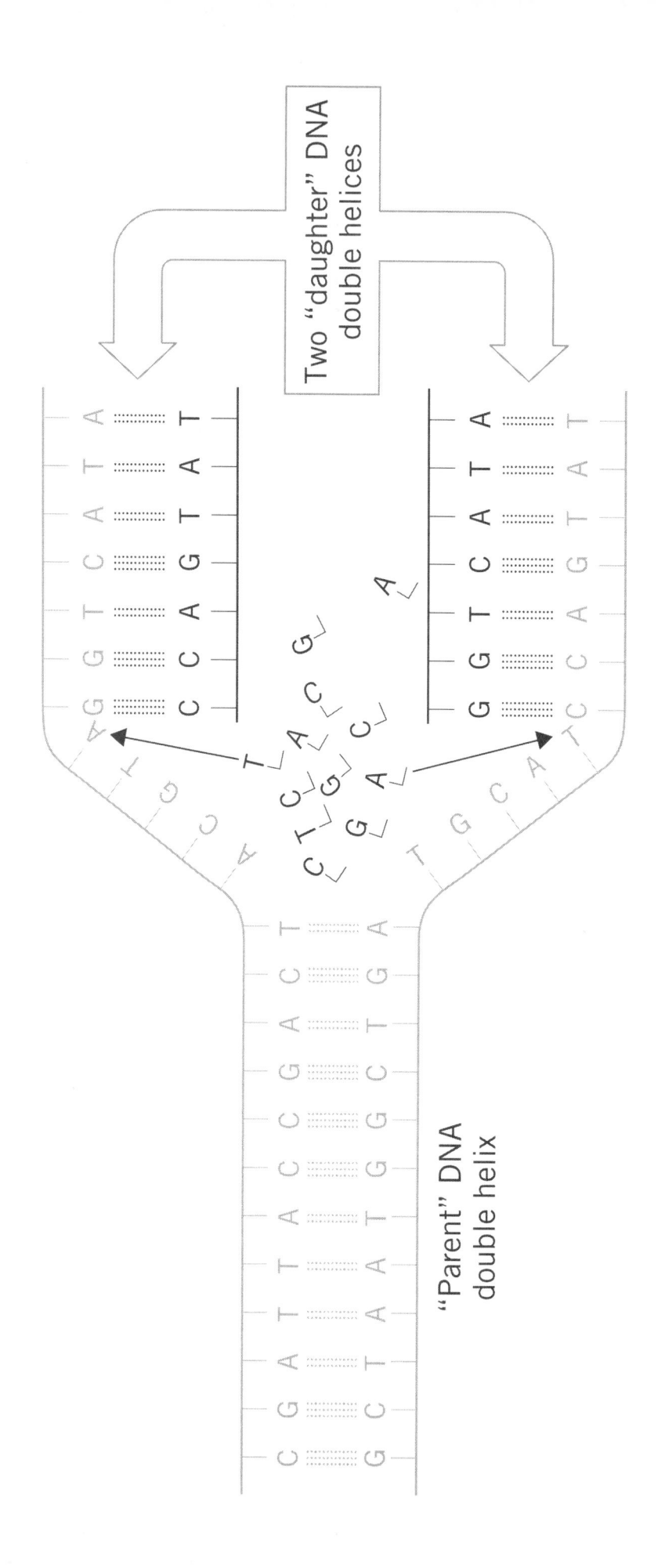

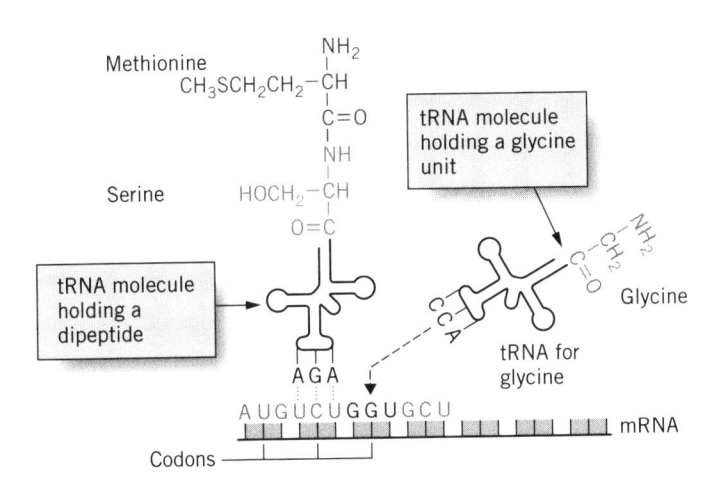

(a) A growing polypeptide is shown here just two amino acid units long. A third amino acid, glycine, carried by its own tRNA, is moving into place at the codon GGU. (Behind the assemblage shown here is a large particle called a ribosome, not shown, that stabilizes the assemblage and catalyzes the reactions.)

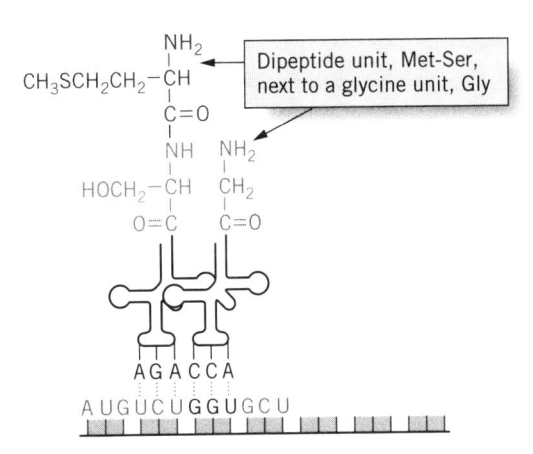

(b) The tRNA–glycine unit is in place, and now the dipeptide will transfer to the glycine, form another peptide bond, and make a tripeptide unit.

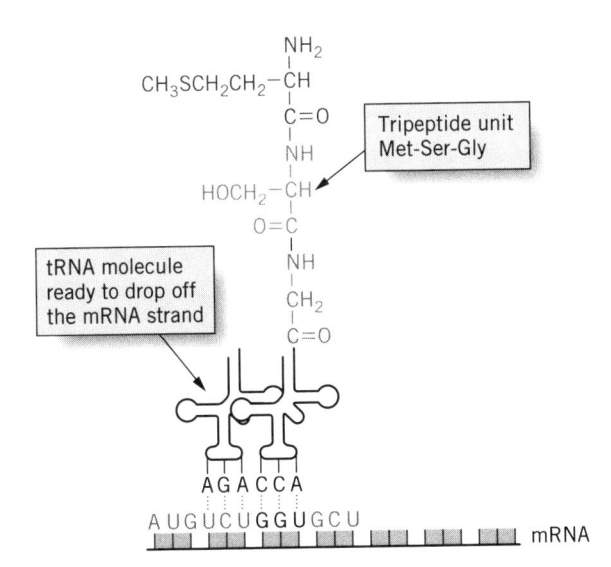

(c) The tripeptide unit has formed, and the next series of reactions that will add another amino acid unit can now start.

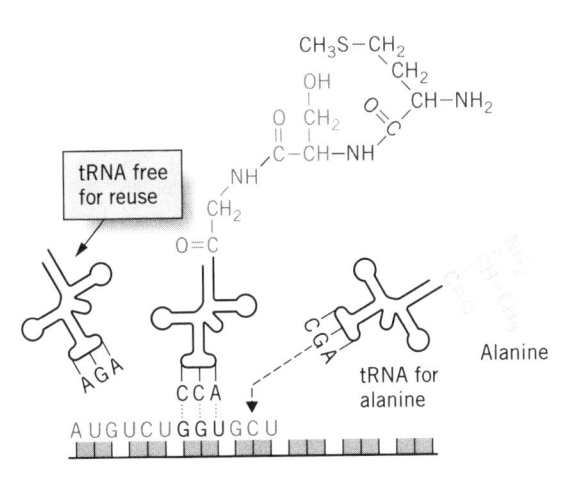

(d) One tRNA leaves to be reused. Another tRNA carrying a fourth amino acid unit, alanine, is moving into place along the mRNA strand to give an assemblage like that shown in part (b). Then the tripeptide unit will transfer to alanine as another peptide bond is made and a tetrapeptide unit forms. A cycle of events thus occurs as the polypeptide forms.

FIGURE 23.15

✍ Take Note!

Take Note!

✍ Take Note!

✎ Take Note!

✍ Take Note!

Take Note!

Take Note!

✍ Take Note!

Take Note!

✍ Take Note!

✎ Take Note!

✍ Take Note!

✎ Take Note!